BIG MOBS

The Story of Australian Cattlemen

> ... this was a romantic period in our history, when unremitting hardships, constant danger and even death were facts of life which were met with both foolhardiness and courage. Perhaps more importantly, they were also met with consummate skill.

Big Mobs is a major social history telling the story of Australia's stockmen. It is a fascinating and detailed insight into the lives of the people who work on the stations and farms, on the stock routes and in the camps. Drawing on both personal accounts and archival research, Glen McLaren has written a rich history of the development of the industry and the technology and skills associated with horsemanship, bushcraft and working cattle.

Glen McLaren was born in York, Western Australia, in 1948. He received a Diploma of Agriculture in 1965 and spent the next ten years developing a farming property, as well as working as a shearer, bulldozing contractor and farm manager. He commenced full-time studies at the University of Western Australia in 1977, graduating with a Bachelor of Science in 1979. While studying he began riding racehorses to keep fit, and this interest grew into a full-time horsebreaking business, which he still operates. Continuing his studies part-time over the ensuing years, he was finally awarded a Doctorate of Philosophy by Curtin University of Technology in 1995.

Glen McLaren's first book, *Beyond Leichhardt: Bushcraft and the Exploration of Australia*, was published by Fremantle Arts Centre Press in 1996. His second book, *Life's Been Good: The Children of the Great Depression*, was published by the Press in 1999. He has recently completed a history of the Northern Territory pastoral industry, and is currently working on histories of the West Australian School of Mines and Muresk Institute of Agriculture.

GLEN McLAREN

BIGmobs
The Story of Australian Cattlemen

in association with the
Research Institute for Cultural Heritage
Curtin University

First published 2000 by
FREMANTLE ARTS CENTRE PRESS
25 Quarry Street, Fremantle
(PO Box 158, North Fremantle 6159)
Western Australia.
www.fremantlepress.com.au

Reprinted 2001, 2007.

Copyright © Glen McLaren, 2000.

This book is copyright. Apart from any fair dealing for the purpose of private study, research, criticism or review, as permitted under the Copyright Act, no part may be reproduced by any process without written permission. Enquiries should be made to the publisher.

Consultant Editor Allan Watson.
Cover Designer Adrienne Zuvela.
Printed by Everbest Printing Company, China.

National Library of Australia
Cataloguing-in-publication data

McLaren, Glen.
Big Mobs: the story of Australian cattlemen.

ISBN 978 1 86368 247 3.
ISBN 1 86368 247 3.

1. Stockmen - Australia - History.
2. Cattle trade - Australia - History. I . Title

636.213092

Publication of this title was assisted by the Commonwealth Government through the Australia Council, its arts funding and advisory body.

Acknowledgements

I thank Curtin University of Technology and the Research Institute for Cultural Heritage for granting me a postdoctoral fellowship to research and write this account of life in the northern cattle industry of yesteryear.

I am also indebted to the Northern Territory Government and the Northern Territory History Awards Committee for their generous grant. Greatly appreciated as well are Mick Muir, for his kind donation, and Henry Esbenshade of the Western Australian Pastoralists and Graziers Association, for his encouragement and introductions.

For their invaluable photographs and information on the northern cattle industry, and for their proofreading and comments, my grateful thanks go to R Bagwell (retired pilot of Perth), Reg Baker (retired horsebreaker and stockman of Toogoolawah), Don Bates (retired saddler of Bates Saddlery, Perth), Edwin Bischoff (saddler and retired rodeo rider of Wandoan), Jennifer Bischoff (of Wandoan), Robert Bradshaw (grazier of Taroom), Frank Dean (grazier of Strathdarr station, via Longreach), Glen Edgar (retired drover and stockman of Wyndham), Annette Henwood (grazier of Fossil Downs station, Fitzroy Crossing), Jim Hill (retired saddler of Syd Hill Saddlery, Dayboro), Noreen Hocking (academic of Curtin University, Perth), Dr Chris Johnson (veterinarian of Epsom Equine Centre, Perth), Dave Ledger (retired horsebreaker and drover of Broome), Darryl Lewis (historian of Darwin, who generously lent me copies of photographs from his Edna Jessop, Marie Mahood and Dick Scobie collections), Lorna McLaren (of Perth), Graham and Angela Moffat (graziers of Camoola Park station, via Longreach), Peter Ross (retired stockman and drover of Fitzroy Crossing), Bruce Simpson (author and retired drover of Caboolture), the Northern Territory University Library, Pat and Peg Underwood (retired graziers of Katherine) and Reg Wilson (retired surveyor of Manton River station, via Darwin.)

Contents

Introduction	9
The Beginnings	16
Mustering	27
Droving	73
Horses and Breeding	113
Horsemanship	138
Saddlery and Harness	168
Horsebreaking	196
Draught Animals	224
Safety and the Use of Firearms	250
Homesteads, Yards and Fences	273
Rangeland Management	297
Equipment and Know-how	317
Epilogue	332
Notes	336
Conversion Table	366
Glossary	367
Index	376

Introduction

In an interesting and provocative article written in 1995, Professor Jim Hoy of Emporia State University, Kansas, claims many Australians tend to see the cowboy of the United States of America rather than the stockman of the northern cattle industry as the 'benchmark for equestrian success'. Hoy bases his assertion on the fact that, whereas he knew very little of Australian rodeos, many Australian stockmen and rodeo identities were conversant with US happenings. Furthermore, he believes most Australians seem 'either a bit defensive or a bit deferential in comparing the [working] cowboy to the stockman.'[1]

His comments, while potentially a rich source of irritation for many Australians, cannot easily be dismissed. Hoy possesses a rare blend of academic knowledge and practical skills, and is perhaps uniquely qualified to comment on the similarities and differences between the two pastoral traditions. Not only has he ridden competitively in United States rodeos, he has also drawn on a family background rich in frontier and pastoral history to produce a series of books on the cowboy and life on the ranges. Furthermore, he has undertaken limited studies of the Australian outback. During a trip to this country in the early 1990s, for instance, he met with R M Williams, rodeo historian Peter Poole and others, examined Australian outback action art, and studied some of our early classics, including Joseph Furphy's *Such is Life*. Thus his perplexity at the deferential attitude of some Australians warrants attention. As Hoy points out:

> In almost all respects I found Australian ranching customs and traditions far more intriguing, more dangerous, more 'western' than those of North America. The heyday of the open-range, trail-driving cowboy in America, for instance, lasted barely a generation ... Not only was the open range a thing of the past by the turn of the century, but the tendency, still continuing as we near the end of the twentieth century, has been to cut large ranches into smaller ones. Big roundups with branding fires were soon replaced by cattle yards and squeeze chutes on the new, smaller ranches, and by the time of World War Two horses were being hauled to the pastures in trailers instead of being ridden there.
>
> But in Australia long cattle drives continued well into the latter half of the twentieth century ... Where an American drover might spend several weeks on the trail, his Australian counterpart often measured his trips in months, even years ... Hostile Aborigines were not only as much a threat Downunder as Indians on the Great Plains, but that threat continued well over half a century longer ...
>
> Ranch size and working methods also favour the notion that Australian stockmen are more rugged than American cowboys ... staying aboard a pitching outlaw in an Australian stock saddle would be, it seems to me, a much greater test of horsemanship than managing

the same feat in an American saddle. Moreover, while roping a wild cow can provide all sorts of tests for bravery ... the Australian practice of tailing down wild bullocks seems much the more dangerous practice.

So why, in light of what seems to me convincing evidence that Australian stockmen are possessed of skills in horsemanship and cattle working that are at least the equal of and in many cases surpass those of the American cowboy, why should the stockman look to the cowboy as his superior?[2]

The answer lies in the fact that the US pastoral experience has been comprehensively documented by historians. Moreover, novelists and artists, and of course Hollywood, have romanticised the 'West', and this blend of fact and fiction has become mythologised. Indeed, according to Hoy:

> popular culture has [capitalised on] our West, making it both accessible to a wide audience and creating a sense of poignance and nostalgia, turning it into a golden age with noble knights roaming an awe-inspiring landscape.[3]

By contrast, Hoy believes that through lack of interest by novelists, artists and historians, 'neither Australian film nor Australian art seems to have [used] ... the Outback experience in the same way.'[4]

While few Australians would disagree with Hoy's premise thus far, many would claim — rightly or otherwise — that the northern pastoral cattle industry[5] does underly some of our myths and legends and typifies much that is 'truly Australian.' Unarguably this was a romantic period in our history, when unremitting hardships, constant danger and even death were facts of life which were met with both foolhardiness and courage. Perhaps more importantly, they were also met with consummate skill. Indeed, whether understood by urban

Australians or not, images of breakaway bulls being turned back to the mob by a hard-riding, stockwhip-wielding horseman are underwritten by the fact that the horse is well trained, the rider can competently employ the lash, and he is capable of pursuing the beast at full gallop through uncleared timber and across dangerously uneven and broken country.

Such skills were a remarkably constant feature of life in the northern cattle industry for almost a hundred years. Certainly, until the mid to late 1940s, the extremely large size of holdings and the unimproved nature of most properties meant that, in large part, station life and work had scarcely changed. Yet, by the 1960s, former stockmen and drovers H M Barker and Matt Savage were adamant that much of the painstakingly acquired cattle sense and everyday skills of that era had already been irretrievably lost. Barker, for instance, asserted that:

> a film of station stockmen trying to yard a mob of cattle in a good many places nowadays would show how far things have slipped. There would be a scared-looking mob of a hundred or so being driven towards the yard gate, whips cracking, men yelling and now and then a bellow from some beast that is being bitten by a heeler dog. Every cow in the mob that has a calf somewhere is either searching for it or trying to protect it in some way and hoping to make a dash with it to get right away. Horses that the four or five men are riding are so upset by bad hands jagging the bit in their mouths that they do not pay any attention to the cattle, so if one beast breaks away the old-time sort of stock horse that would turn and head it in a flash is conspicuously absent. A wild-looking bullock, aged and with long sharp horns, makes a dash away [into the] bush and is out of sight by the time the nearest man has got his horse going. Excessive swearing proves that it is not the man's fault that the bullock got away. That is their idea of it. When the

mob is within fifty yards of the gate, pressure, noise and dust increase, but the cattle do not look a bit like entering the yard. From the back a collection of about a dozen cattle break away and two men set off after them. One gallops on each side of this small lot, so the result is a foregone conclusion; neither can turn the leaders because the other man is forcing them towards him. They give it up and ride back to the main mob which they have left short-handed. They meet the main mob sooner than expected because they are now all off, going full gallop. Nothing can stop them, though a few fruitless attempts are made. Bad horsemanship shows up through the whole performance, the men abuse one another, and two days' work for the gang is lost and much damage done to the cattle as well as the horses.

It would be useless trying to tell these men that fifty years ago, probably on the same station, with the herd, horses and men as they were, and no dogs, two men would have driven the same size mob straight into the yard with no noise other than a few cracks of a whip at any beast that showed signs of planning a breakaway. They just would not understand. Station cattle are getting more out of hand while men and horses become less competent, owners and managers included.[6]

In the thirty years since then the situation has deteriorated considerably. Radical changes have been wrought by the immense development of the road system in northern Australia and the corresponding introduction of road trains, the fencing of properties and the introduction of four-wheel-drive vehicles and two-way radios. The granting of award wages to Aboriginal stockmen, the provision of considerably more watering points and cattle yards (with drafting races and crushes), the quietening of the herd through the frequent musters required under the BTEC scheme,[7] the cutting down of properties and the introduction of aircraft and helicopter mustering have also played a role. Indeed, droving is a thing of the past and helicopter-based contract mustering teams have replaced permanent staff on many properties. Perhaps more tellingly, Heytesbury Pastoral Group, recognising the massive loss of skills which has occurred over the past thirty years, now offers employees TAFE accredited training in such laughably elementary aspects of stockmanship as saddling a horse.

Thus it is vital that this important component of our rich pastoral heritage be adequately documented by historians and cultural heritage professionals before further information is lost. Some progress has already been made in the form of a considerable amount of non-academic writing on the origins and refinement of such skills. Unfortunately these sources of information, though valuable, are frequently less than satisfactory. Often they embarrassingly romanticise the achievements and qualities of men and animals, or are diffuse and tangential. Some exhibit both faults. Such works include nineteenth century Australian explorers' journals, the accounts by British gentlemen such as the Reverend William Haygarth and Captain Peter Cunningham of early nineteenth century bush life, and the late nineteenth century quasi-historical summaries of exploration and settlement by highly proficient bushmen and pioneer graziers such as Edward Palmer and Henry Russell. Turn-of-the-century and early twentieth century examples include reminiscences by Northern Territory customs officer Alfred Searcy and Jeannie Gunn of *We of the Never-Never* fame, as well as the descriptive writings by bushman and author Ion Idriess. Recent works include the memoirs, published in the 1980s and 1990s, of retired drovers, buffalo-shooters, camel teamsters and stockmen such as Tom Cole, Bruce Simpson, Savage and Barker. The few academic writings are generally limited in scope.

This work draws together much of the existing information and corroborates, revises and supplements it with the oral testimony of some of the generation surviving from this colourful era in our history. Given, however, that the major technological developments and writings on the pastoral cattle industry have been contributed by white Anglo-Saxon and Celtic males, there is a strong likelihood this approach will invite criticism. Revisionist historians, for instance, have justifiably documented earlier omissions of Aborigines and women from our colonial history. Certainly no argument is made with claims that some Aboriginal technology might have been appropriated without recognition by the cattle industry, nor that a proportion of those directly involved were women. Nevertheless, it cannot be denied that the pastoral cattle industry has essentially been a masculine pursuit, and that the origin of most of its technology cannot be traced. Consequently, although of passing interest, little emphasis is placed on addressing these concerns. The primary focus of this study is the technology itself, and its chronological and developmental sequence.

The Beginnings

The pastoral cattle industry began, unofficially, when two bulls and four cows escaped their convict herder within a few days of settlement and ran wild in the unlimited bushlands surrounding the fledgeling colony.[8] According to British commentator George Boxall, these cattle, which came from Cape Town, were 'big-boned, slab-sided animals, with enormous horns.'[9] Apart from isolated and accidental contact, they roamed undisturbed for several years.

> A fear of venturing far amongst the natives, then somewhat hostile, repressed all attempts to regain them: indolence succeeded these fears, and no search was ever instituted. [10]

Gradually these cattle moved to and settled in the area later named the Cowpastures, which adjoins the Nepean River. Here they thrived and bred prolifically. Eventually though, in late 1795:

> an officer's servant, shooting in the woods, between twenty and thirty miles from Sydney, discovered them, and conducted the Governor and a party of his friends to the spot, where they found a heard [sic] consisting of nearly sixty head of remarkably fine cattle. The bull attacked the party, who, with some difficulty, escaped unhurt.[11]

Within a further three years the cattle, which were reported to be of 'an uncommon size, and very fierce,' were estimated to number 230.[12]

The earliest official attempt to recapture them was made in 1801, when 'three convicts, who profess themselves equal to the task, ... made an offer [to Governor King] to catch some of them ... by stratagem.' Optimistically, they asked to be paid one beast in ten for the first hundred[13] but, as they lacked horses, stockwhips and practical experience in mustering wild cattle, it is hardly surprising that they and others who followed had little success. Perhaps the best illustration of the complete lack of necessary expertise was the curious finding that a drum beaten in the bush would attract cattle. Clearly this was of no practical advantage, for the mobs of up to three hundred head, having once seen the drummers, inevitably rushed for cover in the gullies and scrub. King acknowledged this technique was of little value when hopefully suggesting it 'may lead to our hitting on some plan to take them hereafter.'[14]

Mustering became more difficult each time the cattle were unsuccessfully pursued, and the task was made no easier when mobs began to reach the broken and scrubby ranges. In a logical attempt to overcome his stockmen's shortcomings, King proposed building a trap on the permanent watercourse at Stone Quarry Creek, in the vicinity of what is now Picton, and using quiet cows as decoys. He admitted, however, that should the plan fail he knew 'of no [other] expedient to take them alive after what has been tried.'[15] Even when sufficient horses became available, the riders lacked the skills essential for mustering in thick scrub and broken country. In 1805, for instance, King was persuaded to surround the cattle 'with a number of horsemen and people on foot [but they] did not succeed in driving any part of them towards the Nepean.'[16]

As the unmanageable herd had little realisable value, both King in 1805 and Governor Bligh in 1807 ordered that the outcast bulls be shot and salted for consumption. Bligh also had as many calves as possible run down and captured alive. While moderately successful numerically, heavy expenses made the operation only marginally viable. The nineteen bulls and one calf shot and eleven calves caught in the 1807 operation were valued at £390, while labour costs and the value of the horse killed by a charging bull were calculated at £307.[17]

In short, all early attempts to control and redomesticate, or at least gain commercial benefit from, this herd, which by 1811 numbered four to five thousand head in the Cowpastures region alone,[18] proved fruitless as the necessary skills and equipment were not available. These problems were replicated on a larger scale once the Blue Mountains were crossed and squatters streamed westward to take advantage of the vast and superb natural grasslands of inland Australia.[19] Now, instead of cleared and fenced agricultural properties, where stock could be managed relatively easily, these new landholders were confronted with extensive, unbounded pastoral runs which were almost devoid of improvements such as buildings, fences, stockyards, roads and clearings. Not surprisingly, their stock ran wild too.

Yet within a very short time these pioneering squatters and graziers began to develop the skills necessary to hold and work stock in a given area. The explorers of the period, such as Assistant Surveyor-General George Evans, Surveyor-General John Oxley and Captain Charles Sturt, also made an important contribution by determining how to traverse unmade country with teams of bullocks and horses, how to find water and how to deal with Aborigines. Their achievements were later rapidly expanded and refined by the first drovers.

Edward John Eyre, the son of an English clergyman, who arrived in Australia in late 1832 as a teenager with limited capital and no bush experience whatsoever, typifies these men.[20] After a brief stint working for a successful and experienced grazier, Eyre went into business, trading in sheep. He rapidly acquired bush experience and skills, so that when within twelve months he decided to take up a property on the Molongolo Plains, near the site of the future Australian Capital Territory, he was capable of moving men, equipment and stock three hundred miles across largely unmarked and unsettled country. Soon afterwards he entered into a lease agreement over three thousand sheep, which had to be transferred to the Liverpool Plains. Eyre encountered considerable difficulties, for the weather was appallingly wet. As a result his drays were continually bogged, and their wheels frequently broke or fell off under the rough conditions. These problems, coupled with his assigned convicts' complete lack of interest in the proceedings, meant that men, equipment and stock were often spread out over unworkable distances. Accordingly, Eyre and his partner were compelled to spend very long hours in the saddle, bringing supplies forward to his disgruntled shepherds and rounding up straying stock, as well as having to extricate bogged wagons. Often:

> After a hard day's work and being wet thro' all the day we had to lay down in the open air at night, frequently in damp clothes and subject to the contingency of the weather which, if not rainy, was bitterly cold. The dews were also very heavy and penetrating. Often we were badly off for provisions, sometimes being for several days at a time without meat ... Each had to cook for himself after his day's work as well as keep watch over the sheep for half the night. Frequently owing to the rains it was impossible to bake even the indigestible bread called damper and the only alternative then was to mix flour and water into a paste and boil it, making what was technically

called 'dough boys', neither a very wholesome nor agreeable way of cooking the staff of life, but still far from unwelcome after a hard day's work ...[21]

This was a hard though valuable apprenticeship for Eyre, and he quickly learnt how to keep large herds of stock together, how to track them if they strayed, and how to cope with unknown and changing conditions. The last factor was of particular importance in 1838 when, after overlanding sheep to the settlement at Port Phillip, he and fifteen men, with two drays, ten horses and fourteen working oxen, set off to drove six hundred cattle and a thousand sheep to Adelaide.[22] At a point well into the north-west of Victoria he found his line of march impossible, for all water supplies had dried up. He reconnoitred extensively in advance for three days but found no water, and was eventually forced to turn back when 110 miles distant from the camp. Eyre soon realised that he would have to punish his horses severely if he was to get back to the camp, and drove them until they could go no further. He then removed their saddles, set them free and, with his men, began to walk the remaining fifty-four miles. Suffering severely from exertion and poor food, they eventually reached their destination, where they took several days to recuperate. Subsequently, the party managed to reach Adelaide via an alternative route.

By the time Eyre completed his overland journey he had been on the track for 147 days and travelled 955 miles. Not only had he further developed and refined the skills necessary to drive livestock over long distances, he and another pioneer drover, Joseph Hawdon, had demonstrated it was possible to move both cattle and sheep from New South Wales to Adelaide.[23]

Within one or two years the overlanding of stock had become an established practice, with large numbers being moved. In his journals of exploration Governor George Grey points out that, in fifteen months during 1839–40, overlanders shifted 11,200 cattle, 230 horses and 60,000 sheep from New South Wales to South Australia. These men quickly learnt to cope with very large outfits of stock, while ideally keeping them fit, sound and thriving. An indication of the size of their operations can be gained from the fact that in one overlander's team there were sixty-two workhorses. Grey, perhaps romantically, described the overlanders as resolute, undaunted, self-confident in difficulties and dangers, and a 'fund of accurate information.'[24]

Certainly, by the end of the 1830s, settlers, explorers and drovers had established many of the features of the pastoral cattle industry. Remote bush life was now an everyday fact. Stockmen had become relatively self-sufficient, more time was being spent in outcamps far away from the homestead facilities, and living and cooking took place primarily out-of-doors. Supplies and equipment were conveyed by packhorses and drays, and the techniques necessary to control livestock without the aid of fences or yards had, to some degree, been developed. Thus the scene was set for the immense expansion of the pastoral cattle industry into the unknown north of Australia.

The pastoral settlement of what was to become the state of Queensland began in June 1840, when Patrick Leslie of Collaroy in New South Wales took up the first station on the Darling Downs. He settled his flock of 'nearly 6,000 sheep, two teams of bullocks and drays, one team of horses and dray, ten saddle-horses and twenty-two … ticket-of-leave men' at the site he named Toolburra, close to the present-day town of Warwick. Although within four years a further twenty-nine properties had been established in the region, expansion was steady rather than dramatic for the following two decades.[25]

In part, this rate of settlement can be attributed to lack of information. In the early 1840s, maps of inland northern Australia were almost virginally white. Cartographers such as Captain Wickham and Lieutenant John Lort Stokes had carried out limited waterborne reconnaissances along major watercourses such as the Victoria, Albert and Flinders rivers. Moreover, Stokes had whetted appetites with his description of the magnificently grassed plains adjoining the Albert River, which he named the Plains of Promise, but little else was known.

This situation changed on 29 March 1846, when the previously little known Prussian explorer Ludwig Leichhardt returned to Sydney as a hero. In a gruelling fourteen and a half months he had covered 2500 miles from Moreton Bay (near Brisbane) to Port Essington, north of the future site of Darwin in the Northern Territory. Leichhardt brought details of extensive grasslands and plains, of a generally interlinking series of watercourses and of permanently running rivers such as the Lynd and Burdekin. According to the early Queensland pioneer Edward Palmer, Leichhardt provided 'the first knowledge we had of the capabilities of North Queensland' and 'his discoveries have been followed by the most extensive and advantageous results.'[26]

Yet, for at least another fifteen years, expansion was slow. Indeed, the rich Peak Downs district, which now supports a variety of grazing and agricultural pursuits, and which was glowingly described by Leichhardt in early 1845, was still only lightly settled in 1857. At that time the Western Australian explorer Augustus Gregory, who was nearing the end of his outstandingly successful North Australian expedition, found the first outpost of settlement there. It was not until the Canoona gold rushes of 1858 and 1859 that pastoral exploration gained impetus. Then, within less than a decade, much of Queensland was taken up and partially stocked.

These must have been heady times, for Palmer speaks of roads being:

> lined with flocks and herds of those entering on the pioneering work of the North of Queensland, and business men [who] were following in the wake of the early stock settlers to commence a trade wherever an opportunity offered.[27]

No time was wasted. Driven by the desire for good grazing country, these pioneer settlers streamed westward and north-westward, continually moving out beyond earlier arrivals. According to historian Noel Loos, within six weeks of George

Dalrymple and his forward party landing at Port Dennison in 1861, settlers pushing inland had taken up all the country for the first 350 miles. Within fourteen months, these pioneers had applied for 454 runs covering 31,504 square miles.[28]

The first cattle, a large mob of five thousand, arrived on the Thompson River in 1862 and at the Flinders and Barcoo rivers in 1864. In the same year the most daring land-seekers reached as far as the Barkly Tableland. George Sutherland, for one, finally settled his mob of eight thousand sheep at Lake Mary. His station, Rocklands, which straddled the Queensland border, is reputed to have been the first property selected and stocked in the Northern Territory, then part of South Australia.[29]

The rapid stocking of new pastoral regions made tremendous demands on drovers, agents and breeders alike. Cattle and sheep were driven very long distances. James Gibson, for instance, set two mobs of cattle under way from the Barwon district of New South Wales in 1861. Passing through Goondiwindi, the Darling Downs, Rockhampton and Bowen, these mobs were then turned inland to Leichhardt's Clarke River. Here they were settled on the first runs taken up in the Burke district and supplemented with other cattle from New South Wales. In 1864, when these properties were sold, the cattle, now numbering in excess of ten thousand, were driven to the lower Flinders region which was then 'quite unoccupied.'[30]

In part, the early Gulf settlers' enthusiasm was underwritten by their firm though mistaken belief that they were ideally located to take advantage of the trade sure to develop, through northern ports, with South-East Asian countries. Nehemiah Bartley, for instance, commented that:

> Normanton, the Gulf port, will be the Singapore of Australia — the great outlet gate of the island continent ... showing by thousands of miles the nearest way to Java, China, India, and all the great markets of the East, for meat, gold and other products ...[31]

Then in 1864–65 a serious drought temporarily checked progress. Conditions were so severe that:

> the native dogs crowded in on the Flinders in thousands, and the blacks themselves had also to resort to it ... None of the rivers ran in their channels and ... stages of thirty or forty miles without water were frequent.[32]

During 1865 and 1866, after conditions had improved, a further group of graziers moved beyond the now partially settled Flinders region and began to settle more closely the Albert, Leichhardt and lower Gregory river basins. At the same time large areas of the central and western plains country of Queensland were being explored, selected and settled. By 1868, however, the outward impetus had lost momentum. The collapse of the British Agra and Masterson Bank, which had underwritten Queensland Government loans, meant credit was frequently revoked.[33] Consequently many of these properties, with their particularly high establishment and running costs, became unviable. Accordingly, stock were moved southward or rendered down locally. Thus, by the late 1860s only 'a few hardy old bushmen held on with grim determination [to their runs in the Gulf region], living on beef and pigweed, and hoping for better times.'[34]

The situation did not stabilise until the Etheridge goldfield, some 250 miles east of Normanton, was discovered and worked. The subsequent discovery of rich alluvial gold deposits along the Palmer River in 1872 rescued the northern pastoral industry. Miners on the Palmer fields consumed an average of fifteen to twenty thousand head of cattle each year[35] and prices soared, for the destocking and abandonment of most Gulf properties meant there were insufficient local cattle to satisfy demand. Speculation in pastoral leases then became rife, leading to the rapid purchase and restocking of abandoned properties, and the resumption of westward expansion.

During the early 1880s large properties on the Daly, Victoria and Ord Rivers were taken up by pioneers such as Fisher and Lyons, the Gordons and the Duracks. These two western regions of immense grasslands, discovered by Augustus Gregory on his 1855–56 North Australian expedition and by Alexander Forrest during his 1878 De Grey to Port Darwin expedition, were stocked with cattle and sheep from Queensland, New South Wales and South Australia. The increase in stock numbers during this expansionary pioneering period was dramatic. In Queensland there was a 730 per cent rise over twenty years, from 432,890 in 1860 to 3,162,752 in 1880.[36]

The initial enthusiasm underlying this unparalleled expansion of settlement soon faded, however. Much of the country so readily taken up, often at ruinous expense, proved unsuitable for sheep. Problems encountered included losses to native dogs, low breeding rates in the Kimberleys and Northern Territory, the cost of freighting wool, the inherent unthriftiness of some of the country,[37] and the difficulty and expense of obtaining shearers and shepherds in isolated and dangerous areas.[38] In addition, during the 1890s large areas of coastal Queensland were found to be heavily infested with liver fluke, intestinal worms and spear grass. The latter was:

> a terrible scourge — [the seeds] ... are finely barbed and intensely sharp and hard; once entered they pass right through the skin of the sheep, even into the flesh, causing great annoyance and leading to poverty and death. [And as a result] ... from Wide Bay to the north scarcely any sheep are now to be met with on coastal runs.[39]

Consequently sheep gradually gave way to cattle over most of the northern pastoral region. Barker notes, for instance, that in 1913, forty-nine years after George Sutherland heroically drove his mob of eight thousand sheep from Rockhampton to Rockland, the 'last of the Barkly sheep enthusiasts also gave it up and changed over to cattle.'[40] The future was similarly bleak for the few hardy west Kimberley pioneers who, in the mid 1930s, were battling dingoes, prohibitive freight costs, grass seeds and blowflies.[41] Overgrazing, poor wool prices and a 33 per cent net advantage to beef producers during the early 1960s meant that by the early 1970s all graziers in the region had converted to cattle.

Mustering

Mustering cattle in the northern pastoral regions of Australia has traditionally been hard, difficult and often dangerous work. In most instances mustering began in late February to early March, as soon as the ground had dried out sufficiently from the wet season to allow horses to move with a degree of safety and ease. Bullocks were usually mustered first and the fat stock suitable for market cut out and held until the drovers arrived.[42] Then, from about May onwards, cows and calves were mustered, marked and branded. Towards the end of the dry season, as the heat increased and feed diminished, mustering came to a halt.

Some wet-season mustering was also done for specific markets. The British firm Vesteys, which had extensive pastoral pursuits in northern Australia, had contracts with the Philippines during the 1920s which called for cattle to be shipped almost all year round. Accordingly, drovers constantly delivered cattle from Vesteys' Victoria River and Kimberley properties to the coastal stations, Burnside and Marrakai, where they were held in bullock paddocks until required. Former public servant Reg Wilson, who continued his very early interest in the pastoral industry before retiring to a small property on the Manton River, south of Darwin, recollects that his father managed two Burnside and one Marrakai wet-season mustering camps. The conditions the men worked under were appalling. Tents, swags, food and firewood were usually sodden, and the men frequently had to swim cattle over flooded creeks as well as gallop across dangerously greasy and boggy ground. Furthermore, whereas 'in the dry season on burnt country you don't have the risk of falling over logs and things,'[43] in the wet season ant-heaps, rocks and logs were hidden in the tall grass.[44]

Prior to the introduction of aerial mustering in the 1960s, cattlemen accepted that a perfect muster could never be guaranteed. There were several reasons for this, including the very large number of cattle involved. Historian Jock Makin notes that in 1934 Victoria River Downs (VRD) station alone was estimated to be running 170,036 cattle,[45] and herds of fifteen to twenty thousand were commonplace. Indeed, the 1959 Queensland beef herd numbered 5.6 million head and that of the Northern Territory 1.01 million.[46]

The immense size of properties magnified the problems of mustering. VRD, which originally covered approximately 12,500 square miles, and neighbouring Wave Hill station were so large that for over twenty years their exact boundaries were neither known nor particularly worried over. When their common boundary was finally surveyed in 1906, Wave Hill was found to be encroaching on VRD by twenty-five miles.[47] While these two properties were clearly the exception, many north Australian stations exceeded two thousand square miles in size. As a result, station herds were commonly spread over very large areas. Indeed, the 1928 Northern Territory cattle herd of 769,000 head was depastured over 232,000 square miles, at an average stocking rate of 3.3 (well hidden) beasts per square mile.[48]

The natural tendency of cattle to disperse into small mobs of up to fifty head compounded these difficulties,[49] while hermit bulls, which had been driven from the mob and roamed alone, were even harder to find and muster. Aborigines exacerbated this problem. During his traverse of the west Kimberleys in the 1930s while accompanying a police patrol, author Ion Idriess spoke with Felix Edgar of Mount Hart station regarding the difficulties of pioneering a station. Edgar listed a number of seemingly insurmountable obstacles, including the rugged terrain, the prohibitive cost of transport, cattle tick and poisonous plants. He also explained that local Aborigines had:

> speared a good many of my cattle, especially down in the Isdell gorges. But it's not so much the beasts they spear that we worry about, it is the constant fear they put into the herds. These split up into small, half-wild mobs ...[50]

The uncleared nature of much of the terrain in pastoral Australia also worked against clean musters. Visibility in areas of central Queensland, the highlands and the Northern Territory is often only one to four hundred yards, and as little as twenty to thirty yards in the brigalow forests of southern Queensland. Broken terrain made the task even harder, for galloping after cattle in very rough country can be extremely dangerous. Wilson, for instance, recalls that bullocks often took advantage of the limestone outcrops that covered much of the ridge country on Burnside and Marrakai. They knew they would not be pursued as vigorously over these outcrops, which are not only very sharp but also partly hidden in the grass.

Providing mustering teams with sufficient horses was a problem for many years, especially when the pastoral, mining and transport industries were expanding rapidly throughout northern Queensland, the Northern Territory and the Kimberleys. Kimberley cattlemen also suffered severe losses from walkabout disease[51] and constantly had to import replacements from South Australia and Queensland. Retired cattleman Pat Underwood, formerly of Inverway station, regularly bought mobs of seventy or eighty horses from Queensland drovers and dealers.

These factors could have been overcome in part if stations had been adequately fenced, for relatively small numbers of unrestrained and constantly roaming cattle are a highly elusive target. Accurate surveying of the boundaries was not necessarily a problem, for Winnecke, Lindsay and others began establishing trig points and carrying out field surveys in what was to become the Northern Territory as early as November 1869. Rather, it was the immense size, low stocking rates and overall poor profitability of northern Australian properties that caused most Northern Territory and Kimberley stations to be still largely unfenced until the 1960s. This situation contrasts strongly with the USA experience. Former cowboy and author Jo Mora writes that, although the prairies were vast, 'the open range was too small ... for so many hundreds of brands shuffling and reshuffling themselves over that unfenced domain.' Consequently the introduction of barbed wire in the late 1870s was a godsend. Not only could ranchers now muster their herds more easily, but they could also upgrade their stock through better control of breeding. [52]

While additional staff might have helped achieve cleaner musters, high costs and low returns made the employment of extra staff unlikely. Paradoxically, although large numbers of poorly paid Aborigines played an important role in the Australian pastoral cattle industry until the introduction of equal wages in the late 1960s, it appears overall to have been understaffed in comparison to the US industry.[53] Cowboy Chas A Siringo, who worked at mustering and droving in that country during the 1870s and 1880s, recalls that prior to the fencing of the prairies a large component of his time was spent in restraining cattle within his employer's boundary. This was particularly necessary in winter, when cattle normally drift-graze with the prevailing wind. Indeed, on one ranch where Siringo worked, cattle:

> drifting southwest ... would have nothing [to impede their progress] but a level plain to travel over for a distance of three hundred miles ...[54]

Consequently, small camps of two men were established twenty miles apart around the forty-mile-square property. Each day these men travelled in opposite directions, half-way to the adjoining camp, looking for tracks that indicated cattle had crossed the boundary. If tracks were found the cattle were pursued and brought back.

The constant herding of mobs on the unfenced prairies also appears to have been commonplace. Siringo mentions being sent out in mid-summer in charge of 'twenty-five hundred steers, a wagon and cook, four riders, and five horses to the man or rider,' and instructed to just drift around the plains 'wherever I felt like, just so I brought the cattle in fat by the time cold weather set in.'[55] Later that season, three more herds arrived and Siringo was promoted to supervise them all. He did nothing but ride from herd to herd, which were up to twenty miles apart, to ensure everything was running satisfactorily and to count the cattle at least once a week.[56]

By contrast, although cattle on new, unfenced Australian stations were herded for some time until they settled into the area, they otherwise rarely saw horsemen except at the annual muster. Certainly there was no consistent application of the practice of tailing, whereby weanlings were drafted off after marking and herded by day and yarded by night for some weeks, to accustom them to horsemen and being collected as a mob. During Matt Savage's eleven years as head stockman at the Montejinnie outcamp on VRD in the 1930s, for instance, the task:

> was too much for a handful of men to control. The cattle were breeding up and all I could hope to do was muster and brand as many of them as possible each year. I could never spare the time to quieten them by

regular handling; if I had tried that I could not have coped with the rest of the work.[57]

Overall, it is hardly surprising that mustering in northern Australia was frequently grossly inadequate, that stations usually carried large numbers of unmarked and unbranded stock, and that in many instances the cattle simply turned wild. Savage's experiences at VRD during the 1930s exemplify these problems. The property:

> was dirty, and likely to remain so. Mustering was always a nightmare. Most of the country was limestone and it was bad [dangerous] to ride about in because of the number of holes. The yards were inadequate and we did not have enough horses.
>
> The bullock paddock had been put up about 1913 and even then it had been a very rough job. By the time I arrived it would no longer hold the cattle. If you put a mob in it one afternoon they would all be gone by the next morning.
>
> In the end I quit Montejinnie because I was sickened by the whole set-up. I had to drive my team of blacks all through the year; yet the job we were trying to do was impossible in the circumstances.[58]

Yet, despite these very considerable difficulties, stockmen developed and refined a number of essential skills, and coped extraordinarily well. At the beginning of each season, horses were mustered and shod, saddles, bridles, ropes, hobbles and bells were repaired or replaced, and provisions packed ready in saddle-bags. Station Aborigines, who were now drifting in from their annual wet season traverse of their country, were accounted for and issued with boots and clothes. Jeannie Gunn provides a sense of the scale, detail and excitement in her description of mustering on Elsey station in the early 1900s.

> We were altogether at the Springs: Dan, the Dandy, the Quiet Stockman, ourselves, every horse-'boy' that could be mustered, a numerous staff of camp 'boys' for the Dandy's work [of yard-building], and an almost complete complement of dogs ... A goodly company all told as we sat among the camp-fires, with our horses clanking through the timber in their hobbles: forty horses and more, pack teams and relays for the whole company and riding hacks, in addition to both stock and camp horses for active mustering.[59]

The techniques employed to find, assemble and control cattle differed markedly, depending on whether the country was open or scrubby. On plains and more open scrub land, the cattle were generally quiet and usually mustered onto waterholes. Accordingly, as many sources of water had to be known as possible.

> To know all the waters of a run is important; for they take the part of fences, keeping the cattle in certain localities; and as cattle must stay within a day's journey or so of water, an unknown water [source] is apt to upset a man's calculations.[60]

Whatever area was chosen, the strategy was usually the same. A camp was established at a major waterhole and all the cattle watering from there were mustered a section at a time. Very early in the morning the musterers commenced riding to the limits of the area they intended covering that day. On reaching that point, they spread out up to half a mile apart and began to drive the cattle in towards the waterhole. Cattle normally begin coming in towards water along the same deeply-worn 'pads' (tracks) by about 10 a.m. By harnessing their natural tendency, a small number of men became far more effective. Cracking stockwhips and bellowing cattle alerted other cattle, which usually moved in the desired direction or gave their position away by bellowing in response. The gradually assembling mob was pushed in towards the waterhole, where it usually arrived just before lunchtime. After a quick lunch and a change of horses, drafting and branding began. Alternatively, where cattle were coming in a long way to water and the distance was too great to cover in one day, skeleton camps were set up. The musterers rode to the periphery of the particular section, carrying only their swags and sufficient food for tea and breakfast. After camping overnight, they got under way very early in the morning.

The approach was basically the same in areas where intermittent ranges formed natural barriers. In the west Kimberleys during the 1930s, Felix Edgar and his Aboriginal stockmen rode up successive valleys, turning small mobs back towards the entrance as they went. At the end of the day's section they turned and worked back, collecting all the small groups and pushing them into a larger mob out on the plain.

While the process was basically one of repetitive searching, it was not an exhaustive combing and re-combing of every piece of ground within a specific segment. Quite apart from this being physically impossible, owners, managers and stockmen knew from experience where cattle could be expected at given seasons and times of the day. Dave Ledger, who managed some of the Emmanuels'[61] properties in the Kimberleys, recalls that:

> you knew where cattle mainly were. We're talking mobs of cattle here, of course ... It's on the country where the sweet feed is, where waters are ... You knew where your cows and calves would be, so you'd mainly dive in there first. You might go back over your diaries and see what you'd normally brand for a season. You'd get in early [in the day] and you'd only have to go out four or five miles. The cattle are all out on the plains or wherever, [and you] round them up pretty quick ... You might have to do two or three musters around that area, but the horses aren't getting knocked around.[62]

Reg Wilson also provides considerable information on the methods his father employed in determining where, when and how to muster in the high-rainfall Top End region. Strategic burning was of particular importance, and in March to April, immediately after the end of the wet season, he would:

> send men to clean around the cattle yards across the runs, and burn the surrounding country. It wouldn't burn all that well in places, unless it had some of last year's [long dry] grass. Come June, when they started the cow and calf muster, they had [fresh green] feed there and the cattle were close handy to the yards.[63]

Furthermore, depending on the season, he knew reasonably well where to search, for large areas of the Marrakai lease are coastal plains and swamps. These areas grow the superb shallow- and deep-water rice grasses, and native swamp grasses. Cattle thrive on these grasses and graze them until forced to higher ground in the wet season by flooding and crocodiles. The latter were a considerable menace to men and stock alike in Wilson's time, for they 'weren't controlled in those days and they were everywhere.'[64] Indeed, on one occasion in 1948 at McKeddy's Hole, a large expanse of water in the Reynolds–Fimiss marshlands, Ray Petherick observed crocodiles attacking cattle trapped on small patches of high ground by rising flood water. As the water advanced and began to inundate these ridges, the terrified, bellowing cattle thrashed in their efforts to clamber on one another. Thus, for some months in the wet season, mustering was restricted to ridge country and higher ground. Even here some areas could be discounted, for, although the limestone ridges grow kangaroo grass and native annual sorghum, the 'needle-sharp' limestone protruding from the ground 'was hard on the cattle's hooves. So they got out of it a bit and onto other dry ridges which didn't hurt their feet.'[65]

Mustering was simplified considerably once bores, mills and dams were developed. This was especially the case on those properties densely covered with scrub. Recalling his experiences inland from Broome, where the scrub was so thick he could scarcely ride through it, author and former Pilbara drover Thomas Cockburn-Campbell mentions that he turned the water off at mills for a day or two. When thirst had forced all the cattle in that area to congregate, he turned the water on again, gave them a drink and then drove them to the nearest yards.[66]

In order to improve the efficiency of musters, particularly prior to the introduction of fencing, sometimes a number of adjoining properties were mustered collectively. On one occasion during the late 1930s, Bulloo Downs, Sylvana, Murramunda, Bald Hill and Cardawan stations in the northern Pilbara region of Western Australia were mustered jointly. Altogether twenty-two men of different 'races' and sixty head of horses were involved, and the branded cattle, cleanskins and calves mustered were returned to the appropriate properties.

A similar practice was widespread in the US before the prairies were fenced. According to Mora:

> In the days of the open range, cows of many brands naturally intermingled and formed a pattern like a well-shuffled deck, and many owners contributed their cowboys to the great round-ups. Large herds were gathered on those fenceless ranges, and at some chosen spot well adapted for the work to be done, the cutting out and branding took place. Steers were separated from cows and calves, and the work of branding the calf crop and the mavericks commenced.[67]

Possibly because of the greater number of men involved, such district musters were very much a social event. Old acquaintances met again, equipment was overhauled, horsebreakers worked on troublesome horses, and conversation, songs and jokes filled the evenings.

While mustering in open country entailed hard work and some fast riding, there is little doubt that it could pall. So later in the season, when the ground was dry and hard, stockmen were rarely dismayed at having to tackle wild cattle. Ledger made a point of varying the work, explaining that by doing so his stockmen would 'have a bit of fun. They'd be throwing [cattle], boasting about what they did. It was good [for them].'[68]

Finding, assembling and controlling wild cattle was an entirely different matter from mustering quiet stock. These animals were shy, timid and very dispersed, and often hid when they realised stockmen were close by. Furthermore, they frequently would not stay in a mob when pursued. Consequently, a lot of hard galloping was involved in an inefficient and time-consuming process which demanded outstanding cattle skills, horsemanship and bushmanship. Three different methods were developed in Australia for mustering wild cattle. The approach used depended on local conditions and how wild the cattle were. In the first, where the scrub was not too thick and the cattle tended to stay as a mob when driven hard, twenty or thirty quiet cattle, known as coachers, were driven out each day to where wild cattle were expected. These were left in the care of two or three stockmen, while a further two or three rode off looking for cattle.

While tracking skills and luck played a part in finding cattle, so too did intuition and local knowledge. When looking for old bullocks on the heavily wooded Glenhaughton station in the Taroom district, contract musterer Robert Bradshaw and his men searched close to the base of ridges and cliffs. Younger, less cautious cattle would be found further out in the valleys. Furthermore, although dogs have not played a big role in the pastoral cattle industry due to the heat and long distances usually encountered, Bradshaw used them on Glenhaughton, where they were very helpful at finding cattle and getting them out of dense scrub and vines. They were also of considerable value when following a rider to the front of a galloping mob and helping turn it. Bradshaw would not allow his dogs to bark, lest they frighten other nearby cattle. He also ensured they did not become too vicious, for he was mustering on contract and the cattle had to be delivered in as good a condition as possible. In Bradshaw's experience big, vicious dogs were a liability. Not only were they injured more frequently through attacking a beast's head, they also tended to tear the tongue, nose, udder or testicles of cattle they attacked, and these wounds frequently became flyblown.

Often musterers would ride for hours before finding fresh tracks, which were then followed, quietly and without speaking, from downwind.

> When it's obvious they are close you have a leak and tighten your girth. As soon as the cattle hear you they are gone of course. You have to try and wheel them and bring them back to the coachers. Sometimes you succeed.[69]

Turning these wild, hard-galloping cattle through thick scrub was difficult, for often they would only 'bend' (turn) once. Any further attempt to bend them would see the mob panic and split. Wilson, for instance, recalls seeing some of the cattle his father was chasing on Burnside in the early 1930s slow down and, seeing their chance, gallop off behind him at right angles.

Once the mob was moving roughly in the correct direction, the musterers, who might have been riding for some hours, had to be able to remember exactly where the coachers were and guide the galloping cattle in that direction. Recalling the exact location was often difficult, as the musterers might have covered ten to fifteen miles in a twisting course, often through flat, heavily wooded and featureless country. Nevertheless, it was not sufficient to come within a hundred yards in thick scrub. There was often only one chance to assemble these maddened beasts and they had to be brought precisely to the coachers.

After the wild cattle had been run in with the coachers, they then had to be settled. Having just been driven at a hard gallop by horsemen and possibly dogs for some miles, they were usually hot and frantically excited. Musterers had to be able to settle them and consolidate the mob before attempting to drive it to the yards. This was achieved by continuously ringing the cattle around in a circle and talking to them. They were then shifted a short distance and re-consolidated a few times. Depending on the time of the day and how many cattle had been found,[70] the musterers often changed duties. Those previously guarding the coachers would ride off looking for cattle, allowing the men and horses that had just worked to have a spell. Towards the end of the day the coachers and mustered cattle would be driven to the yards. Not uncommonly, the musterers arrived at the yards as late as 10 p.m. and the cattle then had to be yarded and watched throughout the night in case they panicked and smashed their way out.

Driving cattle in the desired direction was not always easy. Often they would refuse to be driven and attempt every means possible to break away. Being correctly placed to cut off any breakaways was then particularly important and, as scrub musterer Stan Bischoff of northern New South Wales recalls:

> Old Barry Haigan, he always stayed with the cattle, riding along in front of them on an old mangy black mare. He had a bush in his hand to hunt the flies out of his eyes. Barry never took his eyes off the scrubbers, if one poked its head out of the mob he would be in front of it. They would never get away from Barry.[71]

No one was infallible, though, and when a beast did break away a stockman would ride hard alongside and use his horse to shoulder it around and back into the mob. This could be dangerous.

> You have to put your horse's brisket just behind the beast's shoulder. That way you've got control. If you go up too far, even to their shoulder, they can prop back and open your horse up down the stomach. Bulls aren't too bad, but bullocks and cows are a bit unpredictable. My father had one [horse] opened up

and he took the kangaroo-hide lace out of the end of the reins, and he carried more lace and a needle in his saddlebag. I remember him sewing him up and changing horses.[72]

If shouldering failed, then the unique Australian stockwhip was often used. These whips are of differing lengths, depending on a stockman's personal preference and the situation in which he has to work, for it is difficult to use them in open forest and impossible in heavy scrub country. Although a whip can be plied viciously on a rogue that refuses to cooperate, in most instances cattle have only to hear it crack to comply. Old scrub bulls are notoriously stubborn, however, and if whipped too severely can refuse to budge. Thus it is understandable that, in driving one intransigent beast back to the mob, Charlie Schultz of Humbert River station and his stockmen both shouldered it and 'bashed him over the head with a stick.' Whenever the bull broke out after that, 'everyone would meet him on their horse, [and] come in and yell at him.'[73]

On the occasion when a beast refused to be turned back, rougher treatment was necessary. Australian stockmen used a technique of throwing cattle which involved riding alongside a galloping beast and taking hold of its tail, then turning out at a tangent and pulling its hindquarters around. In most instances this manoeuvre threw the animal off balance, to the extent that it fell on its side or cartwheeled over at full gallop. US cowboys also used this technique, which Mora admits was 'not gentle medicine,' for the bullock would land 'with a thump that would knock the wind out of it completely and even possibly break a horn, a rib, or a neck sometimes.'[74]

Where the animal still refused to cooperate, it was thrown and tied. A number of different techniques were used. In all cases the rider had to catch the animal at the correct gait, a 'half-canter, when he's sort of slightly off-balance and he's getting a bit buggered.'[75] Usually this meant galloping the beast very hard for a short distance and knocking it up quickly. Then, before it got its second wind, it was thrown. The most common method entailed getting:

> the beast in that rolling canter, [then] jump[ing] off your horse and grab[bing] him by the tail. Then race up to his head and he'll turn to charge you. The moment he turns his head around you reef him forward [by the tail] and he'll tip over with his hind legs sticking up in the air and you just wrap your belt around them.[76]

Bischoff's technique was slightly different. He took hold of the bullock's tail while riding and was thus half pulled from the saddle. He would land on his feet and, not having to run and catch the beast's tail, was ideally placed to throw it and tie its legs. This was not common, however, especially in heavy scrub, where the rider could be pulled from the saddle and dashed against a tree.

Other methods included front-legging, at which Northern Territory stockman Joe Dowling was highly proficient. He would run up to a beast's shoulder while it was still cantering at a rolling gait and, grasping its nearside horn with his left hand, simultaneously lean under and catch the offside lower leg. Then, by pushing with his shoulder while the offside leg was clear of the ground, he would tip the beast off balance and throw it. Simpson notes, however, that although this technique relied primarily on skill rather than brawn, it still required a fairly substantial practitioner and he rarely used it.

The disadvantage of these techniques was that if the stockman's timing was wrong and the beast was not unbalanced, it often could not be thrown. Moreover, it might well turn and charge the dismounted stockman before he reached it. On one occasion Charlie Schultz made four attempts to dismount and catch a bull, but stopped each time when it looked back and made clear its intentions. On his fifth attempt, when he finally left his horse, the bull turned while he was still five feet away from its tail. Without hesitating, Schultz 'whipped

Plate 1: The big muster at Keggabilla, north of Goondiwindi, 1930s.

round and ... "off" to the nearest tree as hard as I could go.' Reaching the tree, he hooked his arm around it in order to swing straight around and gain a few seconds' respite. Simultaneously the bull lowered its head, hooked one horn between Schultz's legs and threw him twenty-five feet. Fortunately the bull then charged a jackaroo — who galloped away — while Schultz, mindful of the bull's needle-sharp horns, frantically climbed to safety. He received only severe abrasions to an arm.[77]

Rather than throw a beast by this technique, US cowboys usually galloped behind it and roped it over the horns. They then flicked the rope down either the near or off side before turning out at forty-five degrees to the opposite side and accelerating. This manoeuvre resulted in the rope pulling the beast's hindquarters around until it cartwheeled over. The cow pony then slid to a halt and kept pressure on the rope, while the cowboy leapt from the saddle and tied the shaken and winded animal's legs.

Once thrown, cattle were tied with hobble straps. Musterers usually wore spare hobble straps around their waists and the stockman in white moleskins on the left of the group in Plate 1 can be seen to be wearing two. Young or light animals were tied only on the back legs, while bulls and large bullocks were often tied front and back.[78] Although bigger animals could get to their feet and travel quite quickly when tied, musterers were fairly safe on the ground, for if a beast tried to turn sharply while pursuing them it would fall over. While tied, a beast was sometimes earmarked and castrated, and its horns removed by 'ringbarking [them] ... with a knife around the hairline and breaking them off with a big stick or a stirrup iron.'[79] Alternatively, horns were cut with small pruning saws carried strapped to the saddle.

Tied cattle were left on the ground until the coachers were

Plate 2: Humbert Jack letting up a bullock at the East Baines River, late 1930s.

*Plate 3: Ben McKenzie dehorning a scrub bull while
Stan Bischoff holds its hind leg up, 1939.*

brought up, then released. Occasionally, these infuriated animals would charge the horsemen. Accordingly, some riders selected horses they knew would kick, for a horse can fell a grown bullock by kicking. This strategy was not infallible, however, for sometimes horses became frightened of charging cattle and would only stay and kick those that had been dehorned. In Plate 2, Humbert Jack of Humbert River station can be seen looking pensively at a tied bullock prior to releasing it in the late 1930s.

The second method of mustering wild cattle was used on the 'scrubbers' — those animals that lived permanently in very thick scrub country. Not only were they a continuing source of inferior bulls, but they also attracted quieter station cattle to them. A few highly skilled stockmen, known as scrub-dashers, specialised in catching these cattle. They would gallop after them through dense bush, throw and tie them, and then bring in coachers. Barker recalls that up till World War I the Dawson Valley of southern Queensland was famous for these riders, who worked on a contract basis.

Stan Bischoff of northern New South Wales was renowned as a scrub-dasher. During the 1930s and 1940s he mustered in the Mungle scrub, east of Moree, and in other areas of dense scrub north of Goondiwindi. The Mungle was an area of approximately 50,000 acres of very thick scrub, which, with the arrival of prickly pear, became almost impenetrable. Because the pear was succulent, wild cattle living on it did not need water, so were difficult to entice into open country. Consequently, although Bischoff usually attempted to muster these cattle onto coachers, using dogs when necessary, he also threw and tied a considerable number deep in the scrub. Towards the end of a day's muster, he would return to cattle he

Plate 4: Stan Bischoff with a tied scrub cow, 1939.

Plate 5: Stan Bischoff and musterers, with tie ropes carried around their horses' necks.

had thrown and tied earlier, and dehorn any with dangerous horns (Plate 3). They were then tied to a tree by the head (Plate 4) using the rope carried around the horse's neck. (Plate 5). These cattle would be left for one or two days until Bischoff had time to collect them. Being a bullock teamster, he had large, strong, quiet and well-educated coachers (see Plate 6) and would take out as many of his team as he had tied cattle. Each bullock had a heavy leather neck-strap on, fitted with a hobble chain and swivel, and a spare rope around its neck.

> I would drive one of the bullocks up along side of the [tied] cow, [and] all the gear would be around the bullock's neck. All we would have to do was take the [spare] rope off the bullock's neck and put it through the hobble chain and tie it around the cow's neck, take

the rope off the cow's horns and let them go. They were coupled [close] together, [and] ... the bullock would get her walking [homewards]. By the time I got back in the afternoon the bullocks [and their accompanying scrubbers] would be waiting to get in the gate.[80]

Never one to waste time, after setting the last bullock homeward Bischoff would ride off to catch and tie more scrubbers, ready for the next morning.

Mustering was not all hard galloping and heavy throwing, however, for Bischoff, like Bradshaw and others, was aware that cattle can easily become overstressed and die. Furthermore, he knew that wild cattle living on prickly pear were even less able to cope with stress, to the extent that beasts tied down and left too long 'would be nearly sure to be dead ... when you went back.'[81] He believed this was because their diet was excessively

Plate 6: Stan Bischoff's bullock team at Gladys Downs, 1951.

Plate 7: Stan Bischoff riding through prickly pear at the Mungle Scrub, 1930s.

rich in oil, and referred to the copious amount exuded after death by cattle which had been grazing on pear country.

From experience, Bischoff learnt to select a heavier horse than a thoroughbred type when mustering in prickly pear country. He found that a thoroughbred stallion put to a half-draught mare produced the best cross.

> You would get a heavy legged fellow and if he hit a bunch of Pear [at full gallop], he would knock it down. If you put a light legged horse in the Pear he would probably fall over when he hit a big bunch of Pear.[82] [See Plate 7.]

Most scrub musterers also preferred what they called a scrub-runner. They had found that horses that merely chased cattle were too dangerous in dense scrub, where agility, intelligence and independence were required.

Yet, no matter how skilled the participants, scrub-running was extremely hard and dangerous. In this type of work the rider has to leave the horse's mouth entirely alone.[83] The horse has to choose the path, as it does not have time to await, interpret and respond to commands. Consequently the rider has to sit loose and anticipate which way it will turn and dodge. In his memoirs, Bischoff wrote:

> I [once] had a chap with me by the name of Wally Woods. He was a fair cattleman but a bit mad headed in the scrub. He could catch a beast in the scrub but he used to hit a lot of trees and loose [sic] a lot of skin. In the open going he would go like hell but when he came to a bit of thick scrub he would start pulling his horse's head about and run up trees. When he got through the scrub he would hardly have any clothes on.
> I have always maintained that a good clean rider could have galloped through the scrub and come out without a scratch on him, or a torn shirt.[84]

Even Bischoff suffered injuries. On one occasion he was riding a strange horse, one which had done a lot of scrub-running over the years and was now highly excited at the prospect of a chase.

> When he heard the bullocks ... he took off and of course me not being used to the horse, I must of pulled his head and he ran me into a big Belah tree. Well I don't know how long I was lying at the bottom of the tree, but when I woke up it was well after sundown and there I was — pretty groggy and no horse, right out in the middle of this big scrub.[85]

Two days later, when he had recovered sufficiently, Bischoff rode out to where he had been injured, to find:

the stirrup iron was lying at the bottom of the tree and quite a bit of bark was knocked off the tree, where the horse had hit it. It was all my own fault, because I must have pulled his head and pulled him into the tree. When you are riding in the scrub chasing cattle, you always want to let your horse have his head — if he is any good he won't run you up a tree. When you start pulling your horse's head, nine times out of ten they will run you up a tree.[86]

Understandably, Bischoff was ruthless with horses that dashed him against trees. He would deliberately ride an offending horse headfirst into a big tree to teach it to show more respect for the rider.

Injuries were not restricted to riders. Horses also suffered appallingly. Stake wounds were not uncommon and Bischoff recalls having to tie his chaff-bag saddle-blanket under a horse to hold its entrails in as he walked it some miles to the nearest road. Then, after using a razor blade by the light of a full moon to enlarge the wound, the entrails, along with 'a lot of leaves and dirt and dry blood,' were poked back in and the wound stitched tightly.[87] On another occasion, as Werabone Jack galloped through a big 'melon-hole' (depression in the ground), his horse:

> put her front foot on a stick and it cocked up and went up into her stomach, [and] came out her back behind his saddle. It killed her on the spot.[88]

Even when horses weren't injured, they suffered badly from pear spines. Frequently, after being unsaddled at the end of a day's work, Bischoff's horse would:

> stand there for about an hour maybe, holding one leg up. I used to run my finger up his leg and find all the thorns that were sticking and pull them out, then the pain would go away and he would walk away. Next

Plate 8: Werabone Jack and two other Aboriginal stockmen, second to fourth from left, at a country race meeting.

morning his legs would be all swollen up, but he would not be lame. It would be a couple of days before the swelling would go down, [for] ... the Pear thorns were very poisonous.[89]

Not surprisingly, riders commonly wore knee-length bull-hide leggings to protect their legs from pear. Several of the men shown in Plate 1 (page 42) are wearing such leggings. A former Territory and western Queensland breaker and stockman, Reg Baker, also occasionally saw horses fitted with bull-hide leggings. These were similar to modern-day racehorse leggings, but were in two pieces, above and below the knee.

Bischoff always acknowledged that he had learnt his skills from the Aboriginal musterers who operated from Gooda station, north-west of Goondiwindi in Queensland. These men, who included Joe Brady, Bob Martin, Werabone Jack and Billy

Tindle, were outstanding bushmen and trackers (see Plate 8). When they caught up with wild cattle:

> their idea was to steer them back to where the coachers were. It did not matter if it was a cloudy day or not. Most men, on a cloudy day would get bushed if they could not see the sun — but those black fellows never got lost — they always ran the cattle straight back to the coachers. If any of the men got cut off they would not cooee — they would always howl like a dingo and his mate would answer back in the same way. Their idea for that was if they sang out, like a white fellow would do, they could hunt [scare] all the cattle for miles and they would never catch up to them. Their main idea was to sneak around and not make a noise.[90]

They were also superb riders. As a youth Bischoff and his father accompanied Werabone Jack and the others but, when the group galloped off, they were left hopelessly behind in the dense scrub. Some years later, when Bischoff was highly proficient at scrub-running:

> a chap rode up to the camp on a very poor old horse. He asked Jack if there was any chance of him getting a job with him. It turned out to be Jackey Tommy Tommy, one of the old gun scrub riders, also a full blood aboriginal. He had a bridle on his horse made out of green hide, no saddle, only a bag thrown over the horse's back with a rope thrown over the bag, [and] two mole rings for stirrup irons. He did not look much like a scrub rider. Jack gave him a job and he was the smartest man in the camp. He went right through the muster with us.[91]

It is also clear where Bischoff learnt his technique of throwing cattle.

> When Billy [Tindle] was chasing a bull in the scrub he would race up to the beast and catch it by the tail and then he would let his horse go — the beast would pull him off the horse. There was no running along the ground, trying to catch it by the tail. Billy was with it when he hit the ground and then he would pull it down and strap its legs — it did not matter whether it was a bull or a cow — Billy would get it down.[92]

Indeed, so adept were these men that Bischoff later wrote:

> The aboriginal are the best and smartest cattlemen I ever saw. There isn't a white man who will ever come up to them, especially in the scrub. I was never in the same class as those aboriginals.[93]

The third method of mustering wild cattle was moonlighting. It was used when cattle would come into clearings to graze at night, but not by day. There were some constraints on the process. For one thing, there had to be sufficient moonlight, so only a week of each month was suitable. Moreover, as these cattle were extremely suspicious, the stockmen had to camp well away from the targeted mob. After dark, they drove the coachers to an open area where the wild cattle were known to graze and then hid. Scrub bulls, responding to the lowing of the coachers, would eventually come out and attempt to capture the herd. In doing so, they drew their own herd with them. When the two mobs were well mixed, the stockmen cautiously surrounded them and then rapidly drove them away from the scrub. The coachers, accustomed to being driven where directed, induced the scrubbers to follow. Those scrubbers breaking away were plied with stockwhips, or thrown and tied. All the while the mob was kept circling in order to prevent the wilder scrubbers from all being on the outside.

According to Barker, although moonlighting was a common practice in Queensland in the second half of the nineteenth

Plate 9: One of Stan Bischoff's mustering camps in northern New South Wales, 1930s.

century, it had died out almost completely by 1910. It was not only extremely dangerous for men and horses, but also particularly tiring. Driving the coachers out, capturing the scrubbers and driving them back to the yards could take up to thirty hours. The scrubbers then had to be tailed by day and yarded by night for some weeks to settle them. The main objection, however, was that the scrubbers never really settled and tended to upset the station cattle. Accordingly, Barker claims that with the introduction of military .303 calibre rifles, 'which could hit and kill cattle at twice the distance of any previous gun ... all scrub cattle were stalked and shot and mustering by moonlight ceased.'[94] Bischoff and Wilson provide clear evidence, however, that moonlighting still occurred in the early 1930s. On one occasion Bischoff and a number of men from adjoining properties were mustering Murraculcul station

(Plate 9). Small, semi-independent groups of men were riding off and mustering back onto the thirty coachers, or throwing and tying. One night, to save time, they went moonlighting.

> There was 20 head of scrubbers, [and] these [four] men were throwing and tying them down, while the rest of us were ringing them around so they could not get back into the scrub. The men tied 17 head and three old scrub bulls beat us into the scrub when we were letting them up into the quiet cattle.[95]

Irrespective of the method of mustering, young bulls were often thrown, castrated and earmarked. Traditionally, though, the more dangerous mature bulls were not considered worth the trouble and 'usually found their career cut short by a bullet.'[96] Gunn perhaps best captures the dangers and problems posed by scrub bulls in her highly evocative description of mustering on Elsey station. On this occasion she, her husband, Aeneas, and several stockmen were returning to the homestead. Just as they spotted a mob being driven back to the homestead by station Aborigines, a smaller group of cattle was seen to the side. Aeneas Gunn quickly decided to drive the second mob over to the main herd but, as he and his stockmen galloped forward, their packhorses unexpectedly followed. At the same time it was discovered that many of the smaller mob were scrub bulls which, deigning to be mustered, galloped through the bigger herd. Pandemonium erupted.

> In a moment pack-horses, cattle, riders, bulls, were part of a surging, galloping mass — boys galloping after bulls, and bulls after boys, and the white folk after anything and everything, peppering bulls with revolver-shots (stockwhip having no effect), shouting orders, and striving their utmost to hold the mob; pack and loose horses galloping and kicking as they freed themselves from the hubbub; and the missus scurrying

> here and there on the outskirts of the melee, dodging behind bushes and scrub in her anxiety to avoid both bulls and revolver-shots ... Then in quick succession from all sides of the mob bulls darted out with riders at their heels, or riders shot forward with bulls at their heels, until the mob looked like a great spoked wheel revolving on its own axis. Bull after bull went down before the rifles ...[97]

Even in the 1930s mature bulls still had little value. As Bischoff pragmatically pointed out:

> Old scrub bulls ... were of no use to any one. If you took them to the sale yards you would get nothing for them so they were not worth catching. Dick [Toohey] and I decided to shoot a few of them and skin them, [so] we could take the best hides down to Coonan's Tannery, Toowoomba, to have them tanned. This way [at least] we would always have a bit of good leather on hand.[98]

The economics and desirability of shooting scrub bulls changed dramatically in the early 1960s, however, when North Americans began paying as much for bulls as bullocks — for hamburger meat. Not surprisingly, very few ever got away from Charlie Schultz after that. Once thrown and tied, the bull had its horns sawn off close to the base with a pruning saw carried strapped to the saddle. Schultz did not castrate these mature bulls, for they would lose too much blood and refuse to be driven more than half a mile. Instead he brought coachers over and took them to the yards.

With quiet cattle, once a muster had been completed for the day and the mob assembled, sorting and marking began. Bullocks were usually sorted in the open, after the stockmen had eaten a quick meal from food brought out on the 'dinner-pack.'[99] Then, after the team had changed to fresh horses, a competent rider, usually the head stockman, entered the mob on a camp horse. According to Simpson, camp horses were special horses never used for other work. Quite often they were led out to work and led back afterwards. After selecting a bullock, the rider moved quietly through the mob towards it, then gradually shouldered and cut it towards the outside of the herd.

Considerable cattle sense was required for this work, for it was important that the mob did not become too unsettled and difficult to manage. Several writers have commented that stockmen either had this skill or they did not. Should a bullock attempt to break back, a good rider was hard alongside, quietly anticipating the animal's every move.

> Doubling and turning, twisting and propping, the bullock is determined not to be taken from the mob. But the horse is complete master, one moment blocking it, the next shouldering it until it is forced to the edge of the mob ...[100]

The sophisticated cutting-out process was made easier by correct placement of the cut of selected bullocks. According to Simpson, once a bullock had been pushed clear of the mob:

> the camp riders took over then and took the beast out to the cut. By placing the cut in the direction the cattle usually returned from water, they'd willingly head away from the mob towards their home ground when cut out.[101]

In Plate 10 Ted Fogarty of Delamere station can be seen pushing a bullock away from the mob.

After the selected bullocks were cut out, they were usually yarded for the night and tailed out during the day. As the mob grew beyond a certain size it was frequently watched at night instead of being yarded. Yarding was preferred, however, for watching cattle at night entailed additional tiring and dangerous

Plate 10: Ted Fogarty drafting at Delamere station, August 1923.

work. Accordingly, Charlie Schultz 'put up wire yards through the Upper Wickham [River area of his station], which made things more convenient.'[102] Gunn also notes her husband's continuous program of yard-building on the further reaches of Elsey station.

By contrast, cows and calves were usually taken straight to the yards for branding and marking. Yarding these almost unhandled cattle could be an extremely difficult task. Frequently, very hard riding was called for and it was not uncommon for many of these painstakingly mustered beasts to be lost. Not surprisingly, US cowboys faced the same problems, for, as Mora recalls:

> A crazy critter might put its head down and dash out like greased lightning, and if it couldn't be tailed right there, the riders would let it go, for if they left an unguarded spot for too long, where one got through, a dozen might follow.[103]

Plate 11: Sorting cattle at Dashwood Yard, Victoria River Downs, c. 1940.

Once yarded, cattle had to be sorted, and in the early years this was an unsophisticated process. Often there were only three or four yards, connected by gates manned by stockmen. By opening gates and forcing selected beasts into certain yards, sorting was possible. However, as many of these cattle had had almost no previous handling, and were thoroughly agitated from being pursued, whipped and driven into a confined space, walking among them could be extremely dangerous. Consequently those stockmen who had to go into the yards usually carried a stout pole about three feet long.[104] Plate 11 shows a stockman on the ground using a pole to push cattle towards the gate. In *The Recollections of Geoffry Hamlyn*, written in the early 1860s, the young British-born Sam Buckley is shown drafting cattle by this method. Naturally, as his bride-to-be is present, Sam is portrayed in an intensely romantic light.

Close before her was Sam, hatless, in shirt and breeches only, almost unrecognisable, grimed with sweat, dust, and filth beyond description. He had been nearly horned that morning, and his shirt was torn from his armpit downwards, showing rather more of a lean muscular flank than would have been desirable in a drawing-room. He stood there with his legs wide apart, and a stick about eight feet long and as thick as one's wrist in his hand; while before him, crowded into a corner of the yard, were a mob of infuriated, terrified cattle. As she watched, one tried to push past him and get out of the yard; he stepped aside and let it go. The next instant a lordly young bull tried the same game, but he was 'wanted', so, just as he came nearly abreast of Sam, he received a frightful blow on the nose from the stick, which turned him.[105]

Unfortunately for Sam, his moment of masculine heroism is shortlived.

The maddened beast shaking his head with a roar rushed upon Sam like a thunderbolt, driving him towards the side of the yard. He stepped on one side rapidly, and then tumbled himself bodily through the rails, and fell with his fine brown curls in the dust, right at the feet of poor Alice, who would have screamed, but could not find the voice.[106]

Indeed, carrying a pole did not guarantee safety, and stockmen working in the yards were often charged. If sufficient time was available they usually scaled the yard rails. If not, diving between them was a dangerous option. For this reason yards were frequently constructed with periodic vertical gaps eight to ten inches wide, through which a slim, fit stockman could squeeze quickly. These were known as mangates or manholes.

Plate 12: Scruffing cattle.

After being sorted, stock were branded and marked. Early cattlemen used a roping pole to rope older stock, then looped the end of the rope once or twice around a rail or post to prevent the beast pulling away. The animal was then pulled hard up against the rail and a back leg tied to a post before being branded standing up. Calves were simply caught, bumped off balance and thrown onto their sides in a process known as scruffing (Plate 12). They were then held down and branded. This was often extremely hard work, especially when the calves were well grown. Accordingly, bigger calves were frequently roped and dragged to a rail by a horse pulling from the other side.

Plate 13: Two-man 'parrot-beak' dehorners in use at Victoria River Downs, c. 1940.

Not surprisingly, more sophisticated yards were constructed as stations became better established. Some had races, which are long, narrow laneways constructed of six-foot-high rail fences, for drafting and branding. Races have a gate at each end and are narrow enough to prevent adult cattle from turning around. Once they are loaded with cattle, the rear gate is closed, and branding, earmarking and other operations can be carried out more safely and easily from the outside, or from standing on the rails above. Plate 13 shows stockmen using 'parrot-beak' cutters to dehorn a wildly struggling Victoria River bullock while it is held firmly in a crush at the end of a race.

Although it is not known exactly when races were introduced, there is clear evidence that they were being incorporated in Elsey station's yards by the turn of the century. On one occasion Gunn mentions that newly mustered cows and calves would have to be sent to the Bitter Springs yards to be branded, for the latest yards being constructed were 'not furnished yet with a drafting lane and branding pens.'[107]

While the calf pen and race methods were preferred in much of the Gulf country and coastal Queensland, most cattlemen in the Northern Territory and the channel country of Queensland changed over to branding by the bronco method. The technique of broncoing cattle appears to have been introduced into Australia from Mexico and southern USA in the early 1900s. In this process, the bronco rope is attached to a heavy breastplate[108] or neck collar for, unlike the US roping saddle, the Australian stock saddle does not have a roping horn. (Plates 14 and 15 show both breastplate and neck collar.) A good roper worked quietly through a mob until he was close enough to rope a particular beast (Plate 16). His trained horse then turned of its own accord and dragged the protesting animal towards the bronco panel (Plate 17). As the rider passed the panel, he

Plate 14: Bronco horse with breastplate, Victoria River Downs, 1950.

Plate 15: Bronco horse with neck collar, Victoria River Downs, c. 1940.

Plate 16: Bronco roping at Victoria River Downs, c. 1940.

Plate 17: In the bronco yard at No. 3 Bore, Wave Hill station, 1931.

Plate 18: Broncoing at Victoria River Downs, 1950.

turned so that the rope slid up the angled section and ran along the top of the panel, then fell into the vertical slot. The beast was then dragged further forward until its head was hard against the panel (Plate 18). Here, a team of stockmen threw it by the hind leg and marked and branded it (Plate 19). Those beasts too large to throw were pulled hard alongside the bronco panel and held by leg-ropes attached to pegs protruding six to twelve inches from the ground. All the while, one or two ropers were lassoing and dragging more calves forward. Double-sided panels were used to cater not only for left- and right-handed ropers, but also to increase throughput. When large bulls were to be marked they were double-roped, for, with a horse pulling from either side, the bull could not charge either rider.

Broncoing was usually carried out in yards, where the cattle could not run excessively and become unduly agitated. It was also carried out in the open, where a tree trunk was often substituted for a bronco panel, but open broncoing was very hard on men, horses and cattle (see Plate 20). Charlie Schultz

Plate 19: Bronco branding at Nero Yards, Wave Hill, 1923.

Plate 20: Open broncoing of calves in the channel country, early 1960s.

was scathingly critical of open broncoing, pointing out that, after being castrated, earmarked, branded and let loose, the extremely distraught calf would:

> often ... race straight back to the mob, and it would just as likely keep going out the other side. That would disrupt the other cattle and make it hard for the blokes on horses to hold them ... Cattle would be continually trying to break away and rush, and the horses did nothing but gallop, gallop, gallop, all the time. Talk about a horse-killing game!

Understandably, Schultz commented, 'Jesus, that bloody bronco work used to give me the shits any time, and when it came to open bronco, forget it.'[109] Certainly, by 1959 the practice was viewed with considerable disfavour. In his report on the beef industry in the Leichhardt–Gilbert region of Queensland, J H Kelly wrote:

> Docility of temperament in cattle is an important factor of increased beef production ... A prime cause of nervous temperament is the rough handling of calves and the temperamental upsetting of their mothers by the use of the primitive 'bronco' method of branding calves. This method is gradually giving way to the gentler handling of calves by the use of the branding frame; abandonment of the 'bronco' method everywhere would be of positive economic advantage to beef production.[110]

The method was gradually abandoned as branding crushes were developed and incorporated into station yards.

By contrast, US cowboys used two methods of roping cattle for marking and branding. In the South, the equipment was very similar to that used for broncoing in Australia. However, because a bronco panel was not used and because the cowboys had a much longer tradition of using lariats, more sophisticated roping techniques were used. Smaller calves were mainly roped over the head and dragged to the hardworking branding team, whereas bigger calves were often roped by the hind legs and dragged backwards on their stomachs to the grateful cowboys around the branding fire. In the western regions of the US, cowboys tended to use traditional roping saddles and loop the end of the lariat over their saddle-horn. Bulldoggers were not used in these regions, for calves were roped by the team method. After a cowboy had ridden into the herd and lassoed a calf by the neck, it was dragged out of the mob and brought close to the branding fire. As it struggled and fought the rope, a second roper would lasso its hind legs, loop the lariat around the saddle-horn and back away.

> Needless to say that ... patient was soon stretched out on the ground ready for the operating. In that manner he would be held till everything had been accomplished. Then the ropers would ease up their reatas, shake open the loops, and the calf would scamper to its feet.[111]

While mustering primarily concerned cattle, horses were also involved. Mustering brumbies was somewhat different from mustering cattle for, although they are less dangerous than horned cattle, they are faster than a mounted stockman. Thus, although in his article 'How wild horses are yarded' Banjo Paterson accurately describes how in the earlier years stockmen worked in relays, whereby one rider after another would force the pace and then pull back to allow his horse to regain its wind until the brumbies became exhausted, a more strategic approach was gradually adopted. Crude stockyards were built along the route the brumbies were known to travel. These were usually constructed in the middle of a patch of scrub, or just over a rise, so that the horses would be almost trapped before they realised their predicament. Rail fences known as wings extended out from each side of the gate in order to allow the musterers to hold the mob as they attempted to push them into the yard. Furthermore, bolts of calico were often torn into strips four to six inches wide and hung from bushes in a four- to five-hundred yard unbroken line out from the end of the wings. The fearful and wildly galloping brumbies tended to stay away from the flapping calico and were effectively funnelled into the wings and thence to the yard. On the other hand, a station's permanent yards could also be used. Reg Baker occasionally participated in running brumbies into the homestead yards from a long way out.

By contrast, when mustering brumbies in the rugged Carnarvon Ranges of Queensland, Stan Bischoff frequently galloped alongside the small, poorly grown horses and, lifting them by the tail, threw them in the same manner as cattle. In later years

Plate 21: Stan Bischoff guiding a mob of brumbies being brought down off the Carnarvon Ranges, 1940.

he trapped them in roughly made yards by using salt as a bait. After shooting the old stallions, he side-lined[112] the remainder and brought them down from the ranges (Plate 21).

Overall, in both Australia and the US, mustering was intensely demanding, requiring skills and endurance of the highest order. Both horses and cattle frequently suffered terribly. Galloping through scrub and across broken ground inevitably led to lacerations, stake wounds, bowed tendons, torn ligaments and bone fractures. Over-exertion could also break a horse's wind, so that it was incapable of further vigorous activity. Heat was a potent factor too, and many cattle and some horses suffered from over-exertion and dehydration to the point that they died from heat exhaustion.

Stockmen also suffered. Heat, dust, flies, poor food, extremely hard work, injury and death were facts of life. Boggy country, for instance, could be extremely dangerous for even the most highly skilled.

> The worst thing was that bog. In that black Ti-Tree country there's no bottom in the bloody stuff ... Bog put the wind up you. In that spewy, sandy country, where you saw black Ti-Tree ... or that silver-leafed Ironbark, look out! In a big wet there's no bottom. Some of the poor buggers [the horses] go down to the pommel [of the saddle]. If you jump off and run in front, the poor bastards instinctively lurch in your direction to where they think the good going is, and will pound you into the mud quick![113]

Plate 22: Bogged and exhausted on the Top End coastal plain, c. 1930s.

Plate 22 shows a horse bogged on the Territory coastal plain during the 1930s. For novices the task was appallingly hazardous. For instance, in his early days Siringo roped a very large bullock, which pulled his horse over backwards. As the horse regained its feet Siringo became caught up in the rope, from which he couldn't free himself:

> until relieved by 'Jack' a negro man who was near at hand. I was certainly in a ticklish predicament that time; the pony was wild and there I hung fast to his side with my head down while the steer, which was still fastened to the rope, was making every effort to gore us.[114]

It is hardly surprising, therefore, that some men lost their nerve and became a danger to themselves and to others who were relying on them. As Bischoff recounts, when Tom Toohey and Joe Flood were mustering in the 1930s, they:

> took off after a big old rusty red scrub bull about eight years old. Joe raced up to him and caught hold of his tail and was trying to pull him down but the bull was swinging him around like a tennis ball. Joe was calling out to Tom to come and give him a hand to get him down, but Tom was up a little tree, [and] he said, 'I can't give you a hand Joe, my nerves are gone.'[115]

Droving

In December 1889 Banjo Paterson's poem 'Clancy of the Overflow' was published in the *Bulletin*. While to some extent Paterson was reworking for the public his theme of idyllic bush life in contrast to the misery of city existence, one suspects these lines were written with more than an eye to sales. His portrait of a dingy office lit by a stingy ray of sunlight, of being daunted by pallid-faced, hurrying passers-by and the 'language uninviting of the gutter children fighting,' contrasts strongly with his recollections of the outback, where friends and kindly voices would greet him as he rode towards the 'vision splendid of the sunlit plains extended.' Thus it is small wonder that the nostalgic Paterson wrote:

> In my wild erratic fancy visions came to me of Clancy
> Gone a-droving 'down the Cooper' where the western
>   drovers go;
> As the stock are slowly stringing, Clancy rides behind
>   them singing,

> For the drover's life has pleasures that the townsfolk never know.[116]

While this is a charmingly evocative image, the reality of droving was otherwise. Indeed, few other writers appear to have eulogised the pastime, though many former drovers and observers, both in Australia and the US, have commented on the soul-destroying monotony, the deprivation, hardship, danger and gruelling hard work.[117] The pioneer Queensland grazier Edward Palmer, for instance, states that droving is 'monotonous, wearying and fatiguing in the extreme,' and outlines the early starts, extremely long days in the saddle and hours spent every night watching cattle. Furthermore this grinding workload continued unremittingly, day after day, week after week and month after month, until the mob was delivered and the drovers' responsibility had ended. Palmer also observes that:

> When the weather is fine, life is bearable, if monotonous, but when it rains, especially in cold rain and wind, the pleasures of droving are limited: with wet ground to lie on, wet clothes to ride in, and scarcely fire enough to cook at ...[118]

Campbell develops this, outlining how on one occasion in the Pilbara region of Western Australia during the 1930s rain fell steadily for nearly a week. In an attempt to keep dry he draped a heavy tarpaulin over the cart used to convey food and equipment, and placed the men's wet swags underneath. As he recalls, 'Seven of us were to sleep [there] in wet swags ... We had to eat there too, dashing back and forth to the fire for our food.'[119]

Dust was also a problem for much of the year. Depending on the direction of the wind, drovers might ride through dust for hours each day. Not surprisingly, many were blind by the time they reached seventy from the effects of dust they had encountered over many years.[120]

Another unattractive aspect was that the water available from dams and waterholes was occasionally fouled by cattle trampling, defecating and urinating in it. Moreover, it was not unknown for weak stock to become bogged, die and putrefy in waterholes. Cooks often had to battle against heat, flies and strong winds blowing dust into the food and making the fire difficult to manage. In short, droving was an intensely debilitating occupation for Australian stockmen and US cowboys alike.

Despite the similarities, marked differences can be found in the history of droving in the two countries. A number of factors combined within the USA, from the late 1860s onwards, to ensure ready profitable markets and to facilitate the transport of cattle. At the end of the Civil War large numbers of cattle had been breeding undisturbed in Texas for five years. While the local market was insufficient to absorb this rapidly growing herd, huge markets existed to the north and east. In 1880 the country's population numbered 50,156,000, rose to 75,995,000 in 1900 and totalled 122,775,000 in 1930.[121] Moreover, most northern states, whose herds had been decimated by up to 50 per cent during the Civil War, were willing to pay handsomely for stock. Steers purchased for $3 to $4 in Texas could be sold for up to $40 if transported to the upper Mississippi valley.[122] Unfortunately, these markets could not be tapped immediately, for no recognised major stock routes existed.[123] Accordingly, Texan cattlemen attempted to capitalise on their assets by building rendering, hide and salted beef factories in the Gulf, and by shipping limited numbers of stock to domestic and overseas markets.[124]

This situation changed quickly with three almost concurrent, though separate, developments. Foremost of these was the establishment in the 1860s and 1870s of a network of railway lines across the northern plains. These included the Union Pacific, the Kansas Pacific, the Santa Fe, the Northern Pacific and the Burlington, as well as a number of smaller lines. Secondly, buffalo shooters were wiping out the remnants of the once great buffalo herds at much the same time as American Indians were being driven onto reservations by the United States Army, thus leaving millions of acres of highly fertile soil available for settlement and development. Thirdly, trials demonstrated that Texas longhorns not only could survive a northern winter, but would fatten and thrive in an unprecedented manner in the following spring and summer on the Kansas prairies before being re-sold for slaughter. This was particularly important, for:

> Texas can produce lots of cattle, and grow them up to three or four year olds, but it's hard to fatten a steer on sand, sagebrush and mesquite ... [so] it was the ... Kansas grass — the tallgrass of the Flint Hills, the tall and mixed grass of the Smoky and Gypsum hills, the buffalo grass of the High Plains — that drew the herds from Texas.[125]

Not only did the military, dispossessed Indians, railway workers and new prairie farmers create large local markets, but quick and efficient transport from the fattening properties to the huge northern markets became feasible. As a result attempts were made to develop stock routes, and in 1866 approximately 260,000 head were driven to the railhead at Sedalia in Missouri. This venture was not without severe problems, for heavy rains flooded rivers and bogged the plains, making conditions so unpleasant that many cowboys deserted. As well, Indians deliberately stampeded the herds, and the cattle, 'accustomed to the open range bolted rather than enter [the] forested regions ... of the Ozark plateau.'[126] In the following year, however, Joseph McCoy laid out a new route, known as the Chisholm Trail. It passed primarily through open plains country to Abilene, and in September 1867 the first load of cattle to pass over the Kansas Pacific Railroad was shipped from Abilene to Chicago.[127]

> The cattle boom was on now, and swept over that vast land like an avalanche ... Up the long trails went the longhorns, thousands on thousands of them, to the shipping points springing up on the westward advancing railroads.[128]

Between 1868 and 1871, before the railhead shifted westward to Ellsworth, Newton and Dodge City, a million and a half cattle were loaded in the Abilene yards. Within a few years the Chisholm, Panhandle, Goodnight–Loving and Pecos trails radiated out from the source in Texas, and:

> afforded an exit, a broad highway for the bellowing thousands of longhorns that fed the East for a time and also took the place of buffalo on the great open northern ranges of our West.[129]

With the advent of refrigeration, beef shipments to Europe commenced in 1875 and grew to over seventeen million pounds weight annually by 1884.[130] Thereafter the US cattle industry became frenzied, as eastern farmers and European investors sought to gain a share of the rangelands in Colorado, Montana, Dakota and Nebraska. Collectively:

> Their frantic bidding on cattle to stock the range sent prices skyrocketing; ordinary stock that sold for $8 a head in 1879 brought $35 in Texas by 1882 and were resold in Wyoming at $60.[131]

Thus cattlemen in the United States faced strong and highly remunerative growth from the mid 1860s onwards. Yet droving in that country was short-lived, due to the fencing of the prairies with barbed wire in the late 1880s, the completion of an extensive network of railroads and a high profit margin making rail freight economic. As a result, even:

before the turn of the century north–south railroads made it possible for southwestern cattle to be hauled to summer pasture in Kansas, after which they would be hauled on to feedlots in Iowa or Illinois or to slaughter in Kansas City or Chicago.[132]

By contrast, the Australian industry, particularly in the Northern Territory and the Kimberleys, was affected by limited local markets, poor profitability and transport inadequacies for the first sixty to eighty years. While for some decades the exponential growth in the Australian cattle herd was partially absorbed by the stocking of newer, more distant properties, thereafter profitability was determined by the small domestic market. In 1881, for instance, the Australian population, 'excluding fullblood Aboriginals,' numbered only 2,231,531. By 1900 it had grown to 3,765,339 and by 1930 to 6,500,751.[133] Another adverse factor was the banning of Northern Territory cattle from areas of Western Australia and Queensland in the late nineteenth century, due to the tick-borne redwater fever then prevalent in the Territory. As a result, many properties in the remote Kimberleys, the Northern Territory and the Gulf regions, which incurred heavy establishment, running and droving costs, ran at a loss and were abandoned.

Consequently, well before the turn of the century, cattlemen desperate for some return established rendering works throughout Queensland and the Northern Territory. Local abattoirs, canning factories and refrigerated shipping were introduced later, but were not developed adequately, and in the Kimberleys and the Territory the industry did not thrive until World War II, when huge quantities of beef were required for the large number of military personnel in the north of Australia.

Rail transport could have made a major difference, but the relatively infertile soils and often arid conditions meant turnoff was insufficient to justify a railway system. The only lines to come close to the Territory border were the Toowoomba-to-Cunnamulla line and the Dajarra railhead, south-west of

Cloncurry. The single line to Alice Springs was too far south to be of benefit to most northern cattlemen.

As a result, Northern Territory and Kimberley cattlemen were compelled to drove their stock until the mid to late 1950s, and to this end an extensive network of stock routes was developed. When the Kimberleys and the Territory were first being stocked, the earliest stock routes ran south to north along the general route taken by John McDouall Stuart in his successful 1860–61 crossing of Australia, and east to west along much of Leichhardt's well-watered 1844–45 coastal Port Essington route. What is now known as Elsey station was the junction of these two routes. Two branches then ran from Elsey: one northwards towards the Adelaide River and Darwin, and the second westwards towards the Victoria River district and the east Kimberleys. As the Kimberleys became stocked, the cycle reversed. In this market-driven phase, the length and direction of routes in the Kimberleys and the Territory were at first largely determined by the location of railheads, reliable water supplies and feed, and by ease of travelling. Ranges, for instance, were not crossed if, by travelling a moderate distance further, an easier lowland route could be taken. By contrast, routes in Queensland were mainly gazetted by Government decree and strictly enforced.[134]

As a network of strategically located bores was established in the Territory, a more direct easterly inland route became viable. This route originated from three feeder branches, which came from Gordon Downs, Ord River and Auvergne stations in the Kimberleys. By the time these three branches reached Top Springs, they had merged into the feared Murranji Track. Once clear of the dense Murranji scrub, the route ran eastward to Newcastle Waters, where it was met by the branch coming southwards from Katherine and Daly Waters. These joined, and the main route then ran south-eastward over the Barkly Tableland and along the Georgina River, before crossing the Queensland border at Lake Nash station. While running eastward towards the border, it was joined by a feeder branch

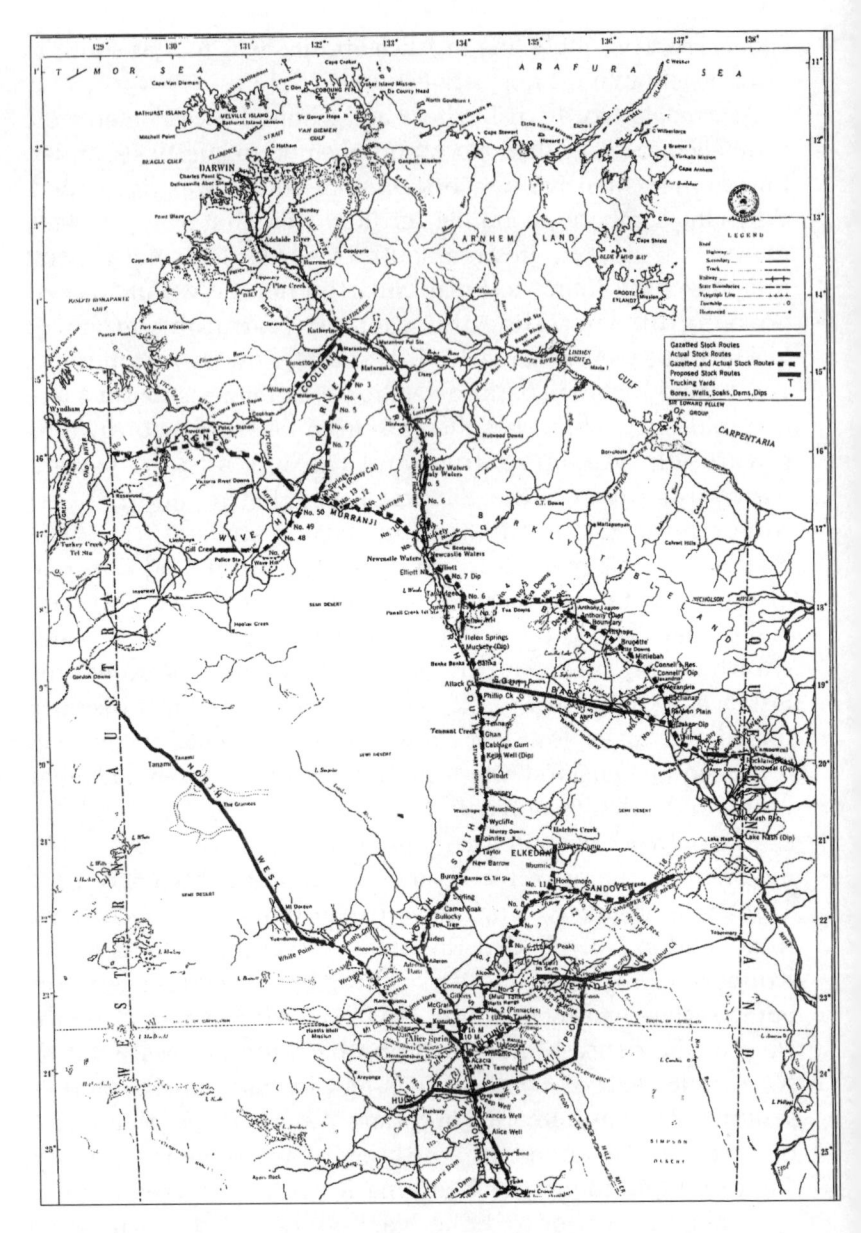

Plate 23: Northern Territory stock routes, 1958.

from the western Gulf region and a further branch from Rockhampton Downs. Finally, a branch left the main route at the Rankine River to cross the Queensland border at Camooweal. By 1958 a number of additional routes had been constructed, as can be seen in Plate 23.

Kimberley cattlemen could also send their stock southward along the Canning Stock Route to the railhead at Wiluna, and thence to the southern markets. Droving cattle across the Great Sandy Desert posed especial difficulties, however, and the Canning route was never a credible option. Cattlemen on fattening properties in the channel country of western Queensland sent large numbers of fat bullocks along the Birdsville Track to the railhead at Marree, from whence they were transported to Adelaide.

Joseph Hawdon, who in 1836 drove three hundred cattle from the Murrumbidgee River to Port Phillip, is credited with being the first to drove stock on a large scale in Australia. At this time all livestock to enter the fledgeling colony of South Australia had to be shipped, usually from Tasmania, which was a slow, inconvenient and expensive process. Consequently the small number of breeding stock within the colony were both insufficient to supply local needs and too valuable to slaughter. Hawdon therefore set out in 1838 from Howlong, near Albury, with a mob of 340 bullocks and followed Sturt's route along the Murray River for almost seven hundred miles before striking across country to Adelaide. He was followed into Adelaide shortly after by Eyre, who had overlanded six hundred head of cattle and a thousand sheep.

These pioneer drovers appear to have grasped the essentials of their trade in a remarkably short time. Indeed, Barker claims that Hawdon was outstandingly successful in what was the first major droving trip in Australian history.

> Hawdon must have been a very good organiser and manager as well, for during the whole of the journey to Adelaide there was apparently no shortage of

Plate 24: Drover's plant, probably Wason Byers', c. 1940.

provisions. That is remarkable. He did not know how long the journey would take or even whether they would get through at all. To take with him everything needed, and yet not have the two drays overloaded, was a masterpiece of planning. Moreover, only four bullocks were lost on the way, and those that reached Adelaide were fat enough to kill two months later.[135]

Nevertheless, they often had to learn from hard experience. There were few precedents and it is understandable that some of their methods seem quaint to latter-day drovers. Learning to use mounted nightwatchmen is a case in point. Initially, mobs were driven into a tight group shortly before dusk, often with a natural barrier such as a steep river bank blocking one or more sides, and the entrapment was completed with a row of fires. The drawbacks of this approach were soon learnt, for the unmounted nightwatchmen had not only to continually stoke the fires, but were also at a serious disadvantage should the cattle break away between the fires during the night. Experiments were also made with different sized mobs in order to determine the optimum size. Experience showed that large herds of four to six thousand were too difficult to handle,[136] while smaller mobs of five to six hundred were found to be prone to rush. Over time, a standard-sized group of 1350 to 1500 bullocks became accepted.

A standard routine was gradually established, in which the boss drover always did the last watch from about 3 a.m. onwards. An hour before daylight he called the cook and the horse-tailer. The cook then woke the men when breakfast was ready. After a hurried meal they saddled their horses from those brought in by the horse-tailer. In a short time the mob moved off in the desired direction, guided by riders on either side near the front, along the sides and bringing up the rear. According to Simpson, the herd stepped out, particularly when they understood the routine and realised they were heading towards water. Drovers aimed at bringing the mob onto water by late morning, after they had covered the bulk of the day's distance.

Plate 25: Camp scene at Victoria River Downs, 1950.

As the cattle were usually strung out over some distance by this time, they were often left to come in and water at their own pace. The horses were then unsaddled and hobbled out, while one stockman kept the tired and usually quiet herd steady. The other men rested as best they could, given the almost ceaseless attention of flies and ants.

Depending on the temperature and the distance to the next camp, after some hours the herd was moved on. The cook and horse-tailer would have gone directly on ahead from the previous camp to set up the night camp.[137] In Plate 24 the horse-tailer is taking part of his 'plant' forward, while Plate 25 shows horses with their packs removed and hobbled out, and the generally spartan conditions. In these last few hours the cattle were allowed to disperse more and drift-graze at their own pace. Just before nightfall the herd was pushed into a mob at the site selected by the boss drover. During the middle of the night all the cattle would rise within a few seconds of each other, relieve themselves, walk a few paces to overcome stiffness, and then settle down ruminating and dozing for a further three or four hours. Shortly before dawn they would begin to stir, then rise and become restless to commence grazing. At this time the drovers had to be ready to move off with them.

Over time drovers became extremely professional, even to the extent of specialising in one of three distinct types of mobs: store cattle, fat bullocks, and cows and calves.[138] Mobs of cows and calves were usually quieter and easier to handle. Edgar recalls that, when droving cows and calves on fenced stock routes in Queensland:

> At night we'd camp across the laneway [to block them breaking back] and someone rode ahead to make sure the fences were alright for five or six miles. Then you let the lead go and a lot of times the next morning you'd have half the stage done before you even caught up with the tail [of the mob].[139]

Driving cows and calves was not a sinecure, however, for they could not be driven as quickly as steers. Furthermore, newly born calves had to be carried in a wagon, known as a calf cart, for a few days until they were strong enough to walk with the herd. Without this conveyance, they had to be destroyed and their deaths caused considerable extra work for the drovers.[140] Fat bullocks were the least trouble, as they were incapable of running as far or as fast as leaner cattle. Moreover, they were generally more placid. In fact, after a few days on the road fat bullocks often did not need watching at night. By contrast, poorly handled lean store cattle, especially those from far-out properties, were frequently extremely nervous and highly dangerous.

Yet, irrespective of the route taken, the size or composition of the mob and the drovers' expertise, serious problems could occur and it is interesting to examine the painstakingly acquired and refined skills that drovers used to counteract them. The greatest problem of all was the ever-present danger of cattle taking fright during the night and galloping wildly through the scrub, for 1500 bullocks, each weighing up to three-quarters of a ton and galloping at twenty miles per hour, generate an immense, unstoppable force. Barker recollects how one night lightning frightened a mob and the cattle began to surge inwards. A second bolt some seconds later showed the cattle in the centre had been forced upwards by the tremendous pressure, and 'one bullock [was] tumbling down over the heads of those leading up to the highest point.'[141] Accordingly, drovers developed numerous techniques for reducing the likelihood of a 'rush', or stampede as it was called in the US. Long distances were sometimes covered at a relatively fast pace for the first few days, especially when the mob was still fresh and unaccustomed to working and staying together. Not only did this tire and settle the cattle, it also quickly removed them from their home range, so reducing the likelihood of their attempting to break away at night. Furthermore, if the cattle were known to be poorly handled and nervous, extra staff were

Plate 26: Counting the mob at handover to drovers, c. 1940.

often employed for the first few days and nights. These were usually the station hands who had mustered the mob and were lent by the station to accompany the drovers to the boundary fence, or just beyond. Plate 26 shows how in the counting-off process, when a herd was handed over to a drover, the cattle were slowly cut out and funnelled down between stockmen, so the manager and drover could count them precisely.

The selection, layout and setting-up of each campsite were also extremely important. Drovers attempted to avoid camping on hollow, drummy ground, where the noise made by a single beast could be sufficient to spook the entire herd. And in scrubby country an area free of fallen or standing timber was selected.

> You don't put 1200 bullocks down in a place where there's a couple of dry trees in the middle, where they can rub up against and the tree will crack.[142]

Ideally, the feed nearby was of reasonable quality, so the herd could be slowly fed up to the selected site and, just before dusk, tightened into a mob and settled down for the night.

The camp was established about fifteen yards back along the track from the mob, leaving just enough room to ride between the fire and the cattle, for cattle attempting to break away usually head in the direction they have come from and thus can often be seen in the campfire light. The campfire was always placed between the mob and the camp. If a wagon was being used to convey supplies, it was located on the opposite side of the fire from the mob and the stockmen slept underneath or behind it. Alternatively, if packhorses were being used, the pack-saddles and swags were placed in the same location. In areas where firewood was scarce and cattle dung was being used for cooking, a hurricane lamp was often hung at the shafts of the cart.[143] In all cases the intention was that, should the cattle rush, the fire or the lantern would deter them and split them to either side of the camp. Simpson, for one, has seen the pack-saddles on either end of the row smashed as a mob split and galloped past. A final point was that it was preferable that a suitable tree was close to the site where the 'kitchen' was located, in order to secure the nighthorses.

By contrast, many stock routes in Queensland had wire yards adjacent to the watering points. By yarding the mob overnight, drovers were freed from much of the worry and hard work of open camping. Some of these yards were avoided, as also were some of the open campsites adjacent to waterholes, because they were notorious for cattle rushing off them. Edgar believes wild pigs were the cause of this problem at one camp near Clermont. Rather than camp at this site, drovers watered their mob, then drove them a further four miles before camping alongside a straight section of fence.

Although camping alongside a fence had advantages, there were also drawbacks. A rushing mob could charge into and through a fence and it was crucially important that the nightwatchman not ride between the mob and a fence. Edgar was caught in the beginnings of this situation one night while riding a big chestnut station thoroughbred, and it 'was galloping before they were galloping. They know. Instinct! He was a good nighthorse.'[144]

Settling the mob down for the night was also important. Nervous bullocks were often difficult to settle and a boss drover usually allowed them:

> room to lie down, or ruminate standing without rubbing shoulders with a restless neighbour, which leaves him little to do beyond riding round occasionally, to keep his 'boys' at their posts, and himself alert and ready for emergencies.[145]

On one occasion Chinese drovers[146] at Elsey station were making a poor fist of holding a mob of bullocks they had just bought. Gunn derived great amusement from their incompetence.

> A Chinaman's idea of watching cattle is to wedge them into a solid body, and hold them huddled together like a mob of frightened sheep, riding incessantly round them and forcing back every beast that looks as though it might extricate itself from the tangle, and galloping after any that do escape with screams of anxiety and impotence.
>
> 'Beck! Beck!' (back), screamed our drovers, as they galloped after escaped beasts, flopping and wobbling and gurgling in their saddles like half-filled water-bags; galloping invariably after the beasts, and thereby inciting them to further galloping. And 'Beck! Beck!' shouted our boys on duty with perfect mimicry of tone, and yells of delight at the impotency of the drovers, galloping always outside the runaways and bending them back into the mob, flopping and wobbling and gurgling in their saddles, until, in the half-light, it was difficult to tell drover from 'boy.' Not detecting the mimicry, the drovers in no way resented it; the more

the boys screamed and galloped in their service, the better pleased they were; while the 'boys' were more than satisfied with their part of the entertainment.[147]

Once the mob had been settled into the camp, the horse-tailer brought the nighthorses up and saddled them. These horses, chosen for their temperament, intelligence and night vision, did no other work. They were ridden around the mob throughout the night watches by the various stockmen, or left tied up in readiness, depending on how nervous the cattle were and how many men were on each watch. Experienced drovers realised when a mob was restless and had more horses saddled ready.

> If you thought something was going to happen you'd have whatever [number] you wanted [saddled up]. But if you thought the cattle were well fed, well watered, and you put them on camp and they sort of plonked down, then you knew you were pretty right. You'd only need two horses.[148]

As a nightwatchman rode around, his horse would see any bullocks attempting to move out from the mob and push them back in. In order not to frighten the cattle, watchmen sang, talked or droned a meaningless chant as they constantly moved through the darkness. As Campbell points out, 'you would get some good entertainment ... and lots of rubbish.'[149]

In short, almost everything possible was done to reduce the likelihood of a rush occurring. Yet they could never be eliminated. The snap of a dry twig or the clink of equipment was often sufficient to frighten one beast. Leaping to its feet, it would startle others and within a few seconds the mob would be galloping wildly through the dark.

> For a few moments the direction of their charge is undecided, but somehow the weight of numbers in one direction decides it for them. Those that are facing

the wrong way in tight formation get carried along until they face the right way, but are very often bowled over, maimed and sometimes killed. The agonising bellow or cry that they give as the others trample over them alarms the whole mob still further and the rush is on.[150]

When this occurred a number of strategies were implemented. Some stockmen ran for the spare nighthorses, while others immediately stoked the fire to deter the cattle and show the riders where the camp was. Not unreasonably, the remainder often scaled a nearby tree. Usually the riders had covered a mile or so before they reached the leading cattle and by this time the mob was often well strung out. Using their whips if the country was open enough, shouting and sometimes even shouldering, they attempted to gradually turn the leaders. Too abrupt a turn increased the risk of cattle further back splitting off and breaking the mob up. Accordingly, a second rider would be backing up close behind. Once the leaders began to bend they were turned increasingly more sharply, until the mob was galloping in a circle. Eventually they would exhaust themselves and slow to a walk, which they might maintain for hours. To steady the herd, the drovers then stayed with them, talking and singing. The agitated mob was then watched closely for the remainder of the night, for once cattle had rushed they were more likely to rush again that night, and for some nights thereafter.

> Sometimes the only sleep you'd get would be on dinner camp, and [even then] them Been Boola mongrels would gallop off dinner camp. A couple of you would get an hour's sleep and that's all you would get.[151]

Surprisingly, few stockmen were killed, despite galloping in the dark through heavy scrub strewn with fallen logs, anthills and rocks, and across creekbeds, and on occasion having to contend with heavy rain. In large part this was because these

Plate 27: Lancewood scrub on the Murranji track, 1930.

were very experienced men, but they were also usually riding outstanding nighthorses.

> I've been in a few rushes, but you'd never miss a beat. You'd just keep going and hope for the best. Your nighthorse would be a proper good horse — a top-class horse. It could see [in the dark] and the main thing is it wouldn't take you underneath trees. It'd swerve out to dodge a tree. You'd be trying to bring it back and he'd be dodging out to get away from the tree rather than wipe you off its back.[152]

Techniques were developed to reduce risk. Charlie Schultz, for instance, timed his departure from his station so he would have full moonlight while crossing one notoriously drummy section of the Murranji track. Not surprisingly, Schultz states that his precaution was based on self-preservation, for the scrub on some sections of the Murranji track was notoriously thick.

Plate 27, dating from 1930, clearly indicates the dangerous nature of this country, and it is understandable that Schultz was fearful of being hit by an overhanging branch.

> The bullwaddy and lancewood doesn't bend, and if you hit a branch it'd knock you clean out of the saddle. When I took off after a beast I'd hold my arm in front of myself so I wouldn't be hit in the head, and I rode loose in the saddle. That way I'd only get it in the hand or arm and it'd be more likely to throw me back onto the haunches of the horse.[153]

Simpson also notes that with two or more stockmen trying to turn a mob in the dark, the danger existed of what he called a 'Chinaman's lane.' By this, he meant riders on either side of the leading cattle unwittingly trying to turn them towards one another. To overcome this, he had his nightwatchmen ride in the same direction, though 180 degrees apart. Thus, if a rush began, one rider would have to go around the back of the mob before he could begin to pursue them, so ensuring both riders were on the same side.

One of the most ridiculous, though humorous, attempts to overcome the problem of nightwatchmen not knowing one another's position occurred in the US when Siringo's trail boss bought new 'bull's-eye lanterns' and insisted his men use them. Within a few hours one cowboy inadvertently flashed his lantern at the herd, spooking them, and 'off they went, as though shot out of a gun.' Siringo immediately galloped in pursuit, trying to head the mob, but he, his horse and the bull's-eye lantern:

> went over an old rail fence — where there had once been a ranch — in a pile. I put the entire blame onto the lamp, the light of which had blinded my horse so that he didn't see the fence.
>
> I wasn't long in picking myself up and mounting

my horse who was standing close by, still trembling from the shock he received. I left the lamp where it lay, swearing vengeance against the use of them around cattle, and dashed off after the flying herd ... [who were] scattered all over the country, badly mixed up with other cattle ... It took us several days to get the lost ones gathered, and the herd in shape again.[154]

Unfortunately, as Siringo indicates, while a rush itself was serious enough, the cattle then had to be found. In the scrublands of pastoral Australia, this could be a difficult and time-consuming task. Drovers did not search blindly, however, for they knew cattle in strange country tend to travel into the prevailing wind and do not mix readily with unknown cattle. Furthermore, because their toes are worn from constant travelling, they leave distinctive tracks. Thus they were able to considerably reduce the area to be searched and, once the correct tracks were sighted, the missing cattle were quickly followed. Stock badly injured during the rush had to be destroyed.

Finding feed and particularly water were also frequently major problems. Early bushmen quickly learnt to evaluate the lie of the land and other indicators of water. Indeed, within three years of arriving in Australia Leichhardt was highly proficient at the task.

> In looking for water, my search was first made in the neighbourhood of hills, ridges and ranges, from which their extent and elevation were most likely to lead me to it, either in beds of creeks, or rivers, or in waterholes, parallel to them. In an open country, there are many indications which a practised eye will readily seize: a cluster of trees of a greener foliage, hollows with luxuriant grass, eagles circling in the air, crows, cockatoos, pigeons (especially before sunset), and the call of Grallina Australis and flocks of little finches, would always attract our attention. The

margins of scrubs were generally provided with chains of [water]holes ... In coming on creeks, it required some experience in the country, to know whether to travel up or down the bed: some being well provided with water immediately at the foot of the range, and others being entirely dry at their upper part, but forming large puddled holes, lower down, in a flat country. From daily experience, we acquired a sort of instinctive feeling as to the course we should adopt, and were seldom wrong in our decisions.[155]

With the advent of large-scale droving, less was left to chance. Alfred Giles, pioneer drover and manager of Springvale station in the Katherine district, records that, while bringing the balance of two thousand cattle and twelve thousand sheep to stock Springvale, Newcastle and Delamere stations, well-sinking parties were working months ahead, making wells and using ploughs and scoops to make waterholes in the riverbeds.

Not uncommonly, even though drovers were following well-established routes, water was unavailable for one or even two days' travel in advance[156] and they had to decide whether to press ahead or seek an alternative route. Time being money, they usually elected to continue on the chosen route. So long as water was guaranteed at given intervals, mobs could be got through. In the nineteenth century, when drovers were still learning what could be expected of cattle, mobs facing one or two waterless days were often watered late in the afternoon, then set under way just before dark. They were walked all through the night and until the middle of the next day, when they were settled onto their dinner camp and rested until dusk, then set off again. By the mid 1900s, however, drovers better understood the capacity of cattle and scarcely departed from their normal procedure. Certainly, 'dry-daying bullocks — that is, watering every second day — was a common practice.'[157] In fact, wells and bores on routes such as the Murranji and the

Plate 28: Charlie Schultz's first mob for 1935 watering at a Murranji bore trough.

Canning in Western Australia were frequently located two days' travel apart.

Once water was found, thirsty stock were watered carefully. Should the supply be insufficient, the troughs too small or the cattle not have drunk for two days or more, the main mob was held well away. Successive small mobs were then cut off and brought in, so that the weaker cattle were not trampled. In critical situations the weakest cattle were brought in first, to ensure they obtained some of the water. Plate 28 indicates the size of bore troughs necessary to adequately water a standard-sized mob of bullocks.

A limitation had to be placed on the size of mobs taken along stock routes where water had to be pulled up from wells. Canning mobs, for instance, were usually kept to five hundred head. The considerable amount of water required was 'whipped up'. Beside each well a large post with a pulley fixed to its top was set into the ground. The process involved passing a rope over the pulley and attaching one end to the harness of a strong whip horse and the other to an eighteen- to twenty-five gallon collapsible canvas bucket. As the horse walked away from the well it raised the bucket to the surface. At a shouted command or whistle it stopped, and the water was tipped into a steel trough some fifty to a hundred feet long by two feet wide. At a further shout the horse backed up slightly, turned around and walked back to the well, thus lowering the bucket. The process was repeated until the mob was watered. Whip horses specialised at this task and rarely did any other work. Camels were also used, especially on the Canning route,[158] although Canning drovers carried engine-driven pumps and fuel on their pack-camels.

Curiously, while stockmen have agreed generally on the indicators of water, such as specific types of trees, there has never been complete agreement on whether animals can smell it. Certainly they have demonstrated a considerable capacity to detect and find water from long distances. At the turn of the century Kalgoorlie prospector and novice bushman A J Macgeorge and his horses were close to perishing from thirst south-west of Coolgardie. He dreamt during the night of swimming again as a schoolboy in the Onkaparinga River, near Adelaide. As he swam he put out his hand to 'caress this water and there was only dry, warm sand. I woke startled and conscious of the peril.'[159] To Macgeorge's despair, his horses had gone, and he desperately followed their aimlessly wandering tracks. Suddenly the tracks went in a straight, purposeful line. After following them for a mile Macgeorge heard the horses' bells and soon found his team alongside a granite depression filled with water.

Similarly, Nehemiah Bartley records how, after thirty-six hours without water, cattle and horses being driven across a stony, waterless forty-mile section inland from Rockhampton suddenly broke into a lumbering trot. They had smelt water still some miles away.

> No flogging or cruel biting at the heels was now needed to keep the stragglers up with the mob. Even

the footsore and lame forgot their pain, and made the pace hotter and hotter as the welcome scent became more distinct ... The rising sun showed the cattle and horses rushing pell mell to quench their burning thirst in a gently flowing creek.[160]

By contrast, bushman and author Joseph Furphy is adamant that 'even wild cattle can no more scent water than we can, though they make better use of such faculties as they possess.'[161] It may be that stock actually detect the scent of damp earth, for Simpson has seen thirsty stock 'ringing madly around a camp when a careless cook knocked over a water canteen, the smell of the wet earth being enough to drive the mob almost crazy.'[162]

While shortages of water demanded certain skills, so too did excesses. Getting a mob into a river was not always difficult, though occasionally the horse plant was driven into the water first to lead the cattle in. Once swimming, the mob had to be carefully controlled to keep them moving in a straight line. Under no circumstances could they be allowed to circle, as otherwise heavy losses could occur. Some stockmen claimed it was important to ride strong swimmers, which allowed them to stay in the saddle and control proceedings. Clydesdale-cross horses, which were renowned for their swimming ability, were favoured. Furphy, in his usual acerbic manner, disagreed, claiming:

> The man who knows no better than to remain in the saddle after his horse has lost bottom, ought never to go out of sight of a bridge. He is the sort of adventurer that is brought to light, a week afterward, per medium of a grappling hook in the hollow of his eye. Perhaps the best plan of all — though no hero of romance could do any such thing — is to hang on to the horse's tail.[163]

Simpson heatedly rejects Furphy's claim, however, with a logical question: 'How the hell can you oversee swimming stock and take corrective action when hanging on to a horse's tail with your eyes at water level?'[164]

Selecting a good landing site was as important as being able to control swimming stock. Without a suitable site the mob could be split up and time would be lost in gathering it together again. Thus a gently shelving bank, up to half a mile downstream from where the drovers intended pushing the cattle into the water, was sought.

Maintaining a horse plant was frequently an extremely worrisome task for drovers, for it was critically important that working horses were adequately fed. Although feed was usually available for several months in the Kimberleys and Northern Territory, where runs were unfenced and drovers could deviate from the route if need be, this was not always the case in Queensland. In many instances the heavily overgrazed five- or ten-chain wide gazetted Queensland stock routes were fenced on both sides and there was little chance of deviating. On unfenced properties, a station hand was commonly sent to ensure the mob kept strictly to the gazetted route and that the statutory daily minimum distance was covered. Whatever the location, as feed declined in quality and quantity towards the end of the dry season, horse-tailers were often compelled to take their horses several miles out from the heavily over-grazed country around waterholes before hobbling, belling and sleeping out with them as they grazed overnight. Throughout the night a tailer would wake, listening for the bells to determine where his horses were. Should they be moving too far away he had to rise even earlier and commence his search. At all costs the plant had to be found, unhobbled and driven in ready for the stockmen as soon as they had finished breakfast.

While finding horses dispersed over a wide area in the dark may seem almost impossible, horse-tailers did not search blindly. They relied primarily on belling a proportion of the horses, usually those known to wander. Because horses mate up, should one or two of a small group be missed when the belled horses were taken, they would almost certainly whinny and give their position away.

Even when the horses had been brought into the camp, the tailer's task was not completed, for in most mustering and droving camps no one else was allowed to catch them. While catching these often poorly handled, nervous horses without the aid of yards was not always easy, the tailer was usually highly proficient and the horses accustomed to his way of approaching them. Each stockman would hand over his bridle and tell the horse-tailer which horse he wanted for the day. The tailer then caught that horse, led it out and handed it to the rider.

> The other way, there's five or six men in there swinging bridles, yelling and screaming, and the whole lot of the horses would [get upset]. A good horsetailer ... well, you never had a crook horsetailer. You couldn't have the bastard.[165]

Undeniably, there is an art in catching horses and cunning animals could quickly take advantage of an incompetent horse-tailer. Furphy captures this beautifully in writing of the hapless British ex-seaman turned stockman whose horses had learnt their superiority and used to 'await, and even approach him, starn-on.'[166] Many horse-tailers caught the quiet horses first and used them to block nervous ones from moving away, in a process which, for the first few days, could take some hours. Occasionally one or two stockmen helped with the blocking, but within a short time most of the horses stood quietly after being brought into the camp and allowed the horse-tailer to catch them. Those known to be difficult were commonly caught in the yards when a fresh plant was being assembled, ridden out and worked on the first day, then hobbled before being let go. Thereafter they were usually sufficiently jaded not to cause problems. By contrast, US cowboys — whose horses were broken to respect a rope — generally surrounded the mob by holding ropes in a circle and lassoed their required horses one at a time.

Curiously enough, there appears to have been a generational difference of opinion in regard to taking in a mustering plant horses that bucked or struck badly, or were hard to catch. Ledger, for instance, claims there were neither the facilities nor the time to cope with horses of this nature and that it was not uncommon for them to be shot.

> The thing is, you get a bad horse, it takes a good man to work it. So he's flat out handling his horse without looking for any cattle. If he gets off it to have a piss, he's got to get back on it again [by himself, which can be difficult] and all this sort of carrying on. They weren't worth it, a bad buckjumper in a plant.[167]

By contrast Simpson, who had finished droving by the time Ledger began, states emphatically that he:

> never heard of a drover shooting a horse. You put the heaviest pack [saddle-bag] on the bastard and, if difficult to catch, you put a shin tapper on as well.[168]

Perhaps the most amusing reference to this ongoing struggle to catch and care for horses can be found in Tom Cole's description of the time Jack Noble arrived at Bullita station in the Territory during the 1930s, mounted on a bicycle. While Cole's veracity has been called into question by a number of elderly Northern Territory residents, his story bears repeating. According to Cole, the arrival of Noble's bicycle was the most sensational event that had ever occurred at Bullita.

> The excitement was indescribable, everyone turned out — my stock camp, all hands, their wives, their children, half a dozen stray nomads and all their dogs. One of the dogs sniffed at the front wheel and started to lift a leg. He only got halfway and an outraged lubra gave him an almighty belt with a

waddy. I thought she'd broken his back at such sacrilege.[169]

Noble explained that he had swapped his two horses, pack-saddles and bags for the bicycle, because:

> All my bloody life I've been getting around the country with horses and packs, finding grass for them, finding water for them and just bloody well finding them every morning ... I thought a bike was a bloody good idea; it doesn't need grass, it doesn't need water and when I wake up in the morning it's lying beside me.[170]

Whatever the approach, considerable skill, ingenuity and perseverance were often displayed in keeping horse plants well fed and ready for work on time each morning.

Conveying food and equipment was the other major problem confronting drovers, who were frequently well away from settled areas and could not always obtain supplies when needed. In 1884, for example, the steamer *Palmerston* reached the Macarthur River loaded with supplies for a station to be established in that region. Members of the ship's crew found the drovers had arrived with the cattle six weeks earlier and, being 'utterly destitute of stores,' had been forced to live exclusively on beef.

> It was fresh beef three times a day — no salt, no tea, no sugar, no flour, nothing but beef. How disgusted those men must have been at the sight of a bullock, either dead or alive.[171]

Even after stores were established close to Leichhardt's Roper and Macarthur river crossing points during the 1880s and 1890s, necessities could not be guaranteed. Indeed, on vast grasslands such as the Barkly Tableland, even firewood was unavailable. Consequently all equipment — which included cooking utensils,

Plate 29: Getting equipment across a flooded Northern Territory creek, c. 1930s.

tin plates and mugs, axes, shovels, branding irons, medical supplies, swags, foodstuffs, a tent and ridge poles, a sturdy, well-sealed water drum, horse shoes and shoeing tools — was carried.

In most instances early Australian and United States drovers carried these supplies in bullock-drawn wagons. The US chuck wagons appear to have been more sophisticated than their Australian counterparts. Mora describes how a large box was often bolted at the tailgate of the former. Its door was hinged at the bottom and opened downwards, where it was held parallel to the ground by chains or ropes. This was the cook's bench and table. Inside the box were various compartments and shelves. Chuck wagons also carried hoops of iron which were erected in wet weather and covered with a tarpaulin.[172]

Surprisingly, even though few river crossings existed in north Australia for the first half century after the country was opened up, transporting these large wagons and equipment across rivers was never an insurmountable problem. As early as the 1820s and 1830s, explorers such as Hovell and Hume, and Surveyor-General Mitchell, took the wheels from their wagons, wrapped the trays in heavy tarpaulins and floated them across. Drovers displayed similar ingenuity when using packhorses. Cole mentions placing his supplies on a tarpaulin and wrapping them into a huge 'plum pudding', which was tied at the top and floated across. Charlie Schultz and Ion Idriess were slightly more sophisticated, using their pack-saddles and tarpaulins to form a boat-like structure, such as that shown in Plate 29, in which they packed their supplies. As Schultz recalls, 'You could carry anything up to a ton-and-a-half like that, no trouble at all.'[173]

Gradually, bullocks were seen as being too slow, for cooks often had to pull ahead of the mob to set up the night camp, or make hurried trips to nearby stations or townships to collect supplies. Accordingly, horses were substituted. While an improvement, they also were vulnerable to wet conditions. Depending on the soil type, many areas became impassable to wagons after moderate to heavy rain. The drover, being paid on a contract rate, could not afford to mark time, so kept the mob

Plate 30: Cook Frank Kearney loading up for another day on the Barkly stock route, c. 1950.

moving forward while the cook and horse-tailer shuttled the necessary supplies and meals forward by packhorses.

Not surprisingly, drovers often changed over entirely to packhorses, despite their inherent shortcomings. Among these was the need to closely balance the weight in the saddle-bags, as otherwise there was a tendency for the saddle to be dragged towards the side carrying the heavier bag. With foodstuffs constantly being consumed, this meant frequent repacking and weighing of the bags. Furthermore, because each bag could carry only a limited volume, a greater number of horses was required. Another factor was the need to saddle and load the horses each morning and unload them at the end of the day, a procedure considerably less efficient than merely unloading and reloading a limited amount of supplies and equipment from a wagon. Plate 30 indicates the amount of equipment that had to be reloaded onto packhorses each morning. Another problem with using packhorses was that tent-poles could not be carried on them. As Barker points out:

> if wet weather comes on and timber of the right sort is not handy for cutting then things become very awkward. Wet weather while droving can be bad enough even with well-rigged tents, but without that its hardships are indescribable. Every man gets either wet through or very nearly so while doing his watch, and comes off watch with feet and fingers numbed with cold. If there is a good tent where he can change into dry clothes and roll into dry blankets then that is a big help, but there is nothing worse than to come into a sagging tent rigged on leather surcingles between two trees with everything wet.[174]

Simpson certainly pared his requirements through years of experience. While conceding that packhorse drovers had a reputation for being mean and that 'the tucker in a packhorse camp was a little short of gourmet standard,' he explains that this was more a matter of necessity. Generally:

> The total load was distributed thus: two packs carrying water canteens, one shoeing pack, one corned beef pack, one flour pack, one cooked tucker pack, and two dry ration packs. The dry ration packs carried sugar, rice, coarse salt for salting beef, coffee, tea, potatoes, curry powder, tobacco, and cream of tartar and soda...
> Cooked tucker pack carried the cook's swag, plus an axe and a small tarpaulin, and a nest of billy cans strapped on top. The rest of the swags went on other packhorses, plus two or three Bedourie camp ovens, a rifle, and a nighthorse peg for open camps [where no trees were close by].[175]

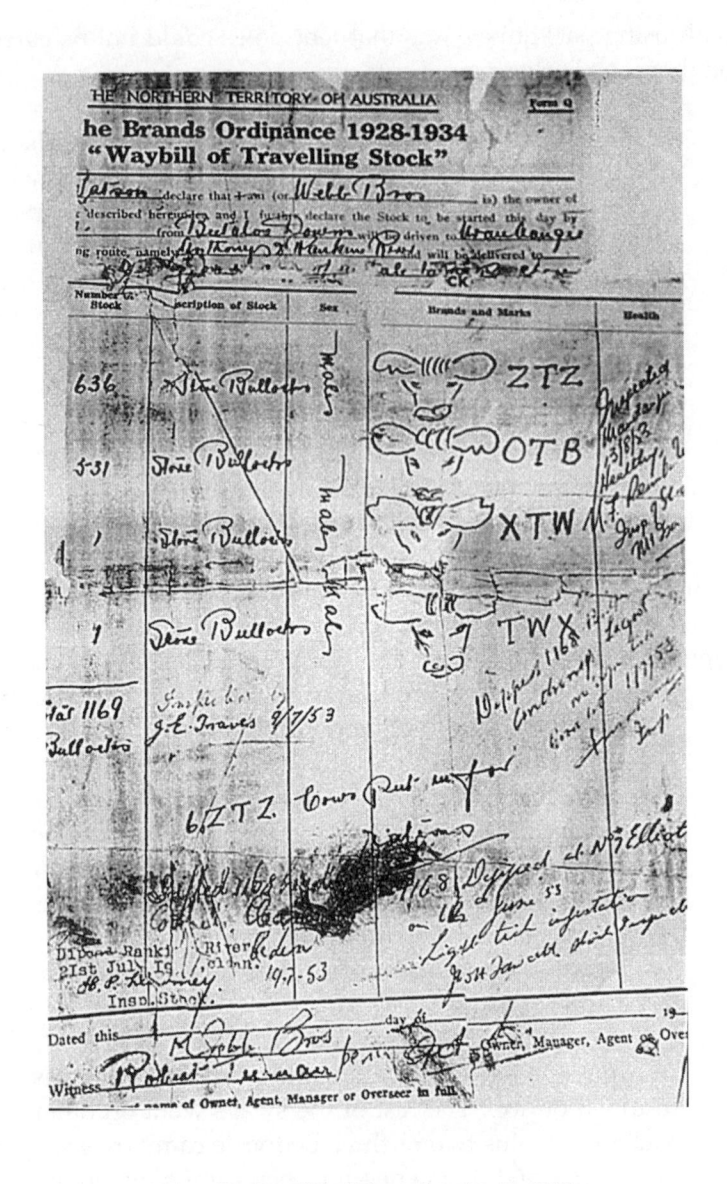

Plate 31: 'Waybill of Travelling Stock', 1953.

Over time, packhorse drovers made minor improvements to their equipment. One of the simplest and most efficient changes was to have one or two saddle-bags made with a wood or iron frame. The outstanding bushman and explorer Ernest Giles used this method as early as 1875, in his successful east-to-west crossing of the Nullabor and Gibson deserts, when he had wooden bases inserted in several of his camel saddle-bags in order to protect his scientific equipment and samples. Similarly, an old, well-worn saddle-bag constructed from a wooden petrol container crate and set hard with greenhide can be seen in the Mount Isa museum. It was donated by Mudgeyacca station of Boulia, where it was used to carry perishables such as bread and damper, which were otherwise squashed when a leather bag was pulled tight by the surcingle strap.

Cattle tick and pleuropneumonia could also cause drovers critical problems. The exact date cattle tick were introduced to Australia is unknown, but officially they were first noticed on Glencoe station in 1880 and attributed to cattle imported from Java. It is more likely, however, that they arrived when buffalo and Timorese cattle were introduced to the settlement at Raffles Bay, north of Darwin, in 1829. These stock ran wild when the settlement was abandoned some time later and probably transmitted the tick when European cattle were introduced into the Territory on a large scale in the early 1880s. As carriers of the bovine disease redwater fever, which frequently decimated herds, tick were responsible for many leases being abandoned. The northern coastal property Burnside, for instance, was wound up by Fisher and Lyons in the late 1880s and again by Vesteys in 1937 because of tick. In an effort to cope with the problem, tick-free zones were established, which cattle from infected areas could not enter unless they had been dipped in an arsenical solution, then inspected and declared tick-free. These restrictions and inspections were administered initially by the Northern Territory, Queensland and Western Australian mounted police, and later by government stock inspectors. The stock inspectors' signatures on the Waybill of Travelling Stock

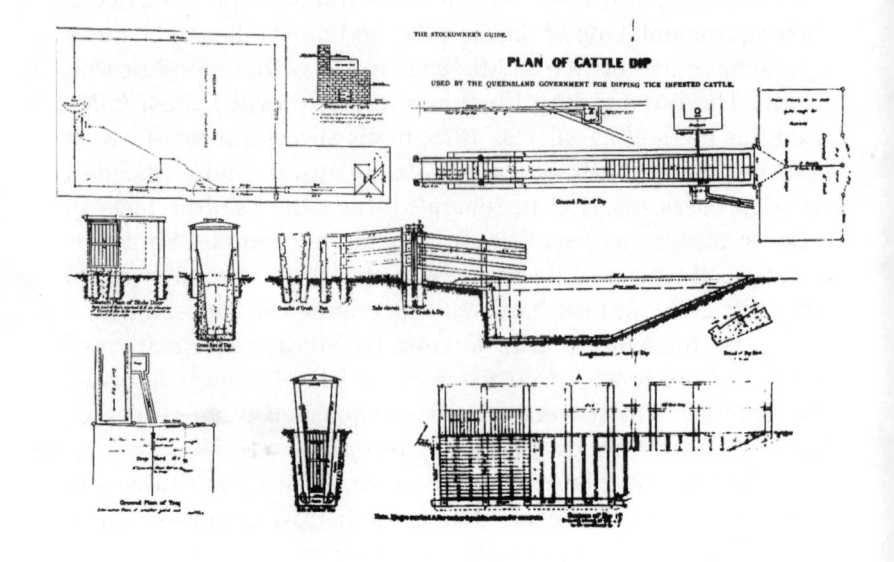

Plate 32: Plans of a Queensland government-approved cattle dip.

shown in Plate 31 indicate that the mob referred to was dipped three times and inspected twice in nine weeks.

A government dip was established at Lake Nash on the Queensland–Territory border in 1907 but, when it became apparent that more than one dipping was often required, additional dips were constructed at Austral Downs and the Rankine in 1918. Others were built later at Anthony's Lagoon and Camooweal. These allowed drovers to keep moving and also reduced the grazing pressure around the original dipping point. These dips were incorporated within the station yards and included forcing races, draining pens, dip-mixing and topping-up tanks, and cement-lined, corrugated-iron-roofed dips. Plate 32 indicates the scale and sophistication of the dips recommended by the Queensland Government, while Plate 33, dating from 1931, shows the first mob of cattle to be dipped on Wave Hill station.

Plate 33: First cattle dipped at Wave Hill station, ex Willeroo, 30 September 1931.

In their contracts, drovers were allowed a day's travelling time for each dipping. The work involved in yarding, penning and forcing these often almost wild cattle through a plunge dip was immense. Even when all went well, they often became agitated and prone to rush for some time after, but if the dip solution was too strong, the scalded cattle were almost unmanageable. Cattlemen learnt to use washing soda in the mix in place of caustic soda, for it was less likely to scald and irritate the animals. The amount of soda used varied according to the properties of the local water and cattlemen quickly understood the desirability of dipping only a few cattle at first.

> If [after dipping] the test cattle are restless, and off their feed, the probability is that the mixture is too strong. If, on the other hand, the cattle are happy and some of the ticks still alive, the dip is not strong enough. When the quantities are right all the ticks should be dead next day, and the beasts be about normal.[176]

As the arsenic soon stripped out from the beast's hair, Stockholm tar and tallow were sometimes included in the solution to bind it more strongly.

Contagious bovine pleuropneumonia, or 'pleuro' as it was commonly called in Australia, was known in older literature as lungsickness. It originated in central Europe and spread throughout the Continent during the Napoleonic Wars. By the early 1840s it had crossed the channel to Britain, was recorded in South Africa in 1854 and reached Australia in 1858. The ability of the virus to survive the sea voyage was almost certainly due to its long incubation period, which could be as much as 207 days. The virus is spread by droplets in expired air, and thus infection depends on close contact. Pathological changes include inflamed and swollen lungs, the development of oedemal pleural fluid and eventually destruction of lung tissue. In the acute stage:

the animal is in extreme distress with laboured respiration, heaving flanks ... dilated nostrils and drooling and frothing at the mouth, and is ... unwilling to move.[177]

Effectively, beasts with the disease choke to death and losses of up to 50 per cent of a mob were common.

In order to prevent travelling stock spreading the disease, every animal had to be inoculated and bang-tailed before a drover took delivery of a mob.[178] Attenuated vaccines were developed from immature lymph fluid taken from infected animals but, due to the variability of the virus, this frequently induced severe reactions. Former Territory cattleman Felix Schmidt of Alroy Downs later wrote, 'Pleuro Pneumonia in cattle was the scourge in droving mobs from the N.T. for so many years.'[179]

While the bulk of animals moved across country were cattle, the droving of horses also became an important business. For many years large numbers of horses were required for the rapidly developing stations in northern Australia. The Kimberleys, where walkabout disease ravaged horse herds for nearly a century, had a particular need. Barker mentions a station on the Cape River, east of Charters Towers, which ran ten thousand horses in the early 1900s, from which the owner sent mobs of five hundred to the east Kimberleys under a drover named Heron.[180]

With their capacity to gallop faster and further than cattle, horses posed special problems for drovers, and stockmen soon realised that these animals needed to be allowed to stride along faster than cattle. Whereas the latter would usually cover only ten to fifteen miles per day, horses would travel more than twenty-five. Moreover, the front of the mob had to be kept under control, lest they begin to gallop and take the others with them. Accordingly, two or three stockmen would ride in front of the mob, blocking and steadying it by talking to the horses. As Aboriginal labour was usually plentiful, the mob was generally closely guarded. Surprisingly few problems were encountered during the nights, for horses seem less prone to rush than cattle. It was always a possibility, however, and most drovers mixed their quiet working horses among the unbroken ones. As these animals were hobbled, they were a steadying influence on the mob. A good number were also belled and one or two stockmen rode on watch throughout the night. After a time the mob would settle and, while 'every now and then one of them would poke his head out and have a bit of a pick,'[181] many drovers considered horses easier to manage than cattle.

Yet despite this carefully developed expertise, and the establishment of a comprehensive network of stock routes and strategically situated water supplies, the end of droving came relatively quickly. As the road system was expanded dramatically throughout the north of Australia during World War II, and in response to the development of major mineral projects in subsequent years, the large-scale trucking of cattle became both possible and viable. The major benefit of trucking was that stock arrived at their destination within a day or so of being loaded, thus avoiding the loss of condition usually associated with droving. It is therefore hardly surprising that, although in the early 1960s approximately 60 per cent of Northern Territory cattle were shifted by drovers, the last mob went across the Murranji in 1967. By the late 1960s more than 90 per cent of Territory cattle sent to market were transported by road. Since then many bores, windmills, tanks and troughs having fallen into disrepair, and droving as it was traditionally carried out is no longer possible.

Horses and Breeding

Horses have played a vital role in the Australian pastoral cattle industry. Without them, the delivery of supplies would have been far more difficult, and travel slow, expensive and often dangerous. More importantly, as the experiences of the earliest cattlemen showed, the herd would have been unmanageable. Yet the development of the Australian horse herd was a slow process for several decades. Perhaps the most significant factor inhibiting early growth was losses in transit. When the First Fleet called for supplies at the Cape of Good Hope en route to Botany Bay, it also took on board four mares, one stallion, one colt and two fillies, as well as sheep, cattle, goats, poultry and fodder. The fleet sailed from the Cape on 11 November 1787 and disembarked the male convicts, marines and livestock at Sydney Cove on 28 January 1788. Only six horses were unloaded, the fillies having died in transit.[182] Similar losses occurred in later voyages, with the brig *Britannia*, for instance, losing ten out of forty-one horses in 1795.[183] Necessity also affected growth of the herd, for the colony urgently needed to be self-sufficient in meat, and the importation of breeding cattle and other livestock was given precedence over horse breeding.[184]

The cost of importation also markedly inhibited growth of the herd. A 1797 quotation for the shipment of cattle from the Cape of Good Hope to the Colony — fodder, insurance and handling included — was £35 per head.[185] As the typical wage for a labourer in 1797 was fifteen shillings per week, it is clear that the presumably comparable cartage cost alone for a horse was almost equivalent to a year's wages. In 1999 values, that would be at least $25,000. Thus by 1796 there were only fifty-seven horses in the colony, all of which were owned by the Government or by civil and military authorities, and ten years after settlement there were still only 117.

By 1806 the herd had increased to 552 and, apart from one stallion and one mare brought from Britain in 1802,[186] they had all either originated in or been bred from stock imported from the Cape or India. In most cases these horses had some infusion of Arabian blood. Mares were subsequently also imported from 'Valparaiso, and pony mares from Lombok and Timor.'[187] At this time horses were extremely expensive, ranging in price from £100 to £150 for ordinary mares from 13.2 to 15 hands high.[188]

Figures 1 and 2 examine the availability and cost of horses from the time of first settlement to the mid 1840s. Figure 1 indicates the growth of the European population relative to the horse population. Because very small numbers of horses were involved in the early years, approximate human–horse ratios have also been included. Figure 2 indicates that horses were at their most costly in about 1806, though it has to be noted that this observation is based on very meagre data. Perhaps the most useful indicator of the cost of horses is, once again, the price relative to a labourer's wage.[189] In 1800, at £50 per horse and fifteen shillings per week wage, a horse's value would have been approximately equal to sixty-seven weeks' wages, in 1806 a hundred and twenty-six weeks, in 1825 fifty weeks and by 1837

Approximate ratios

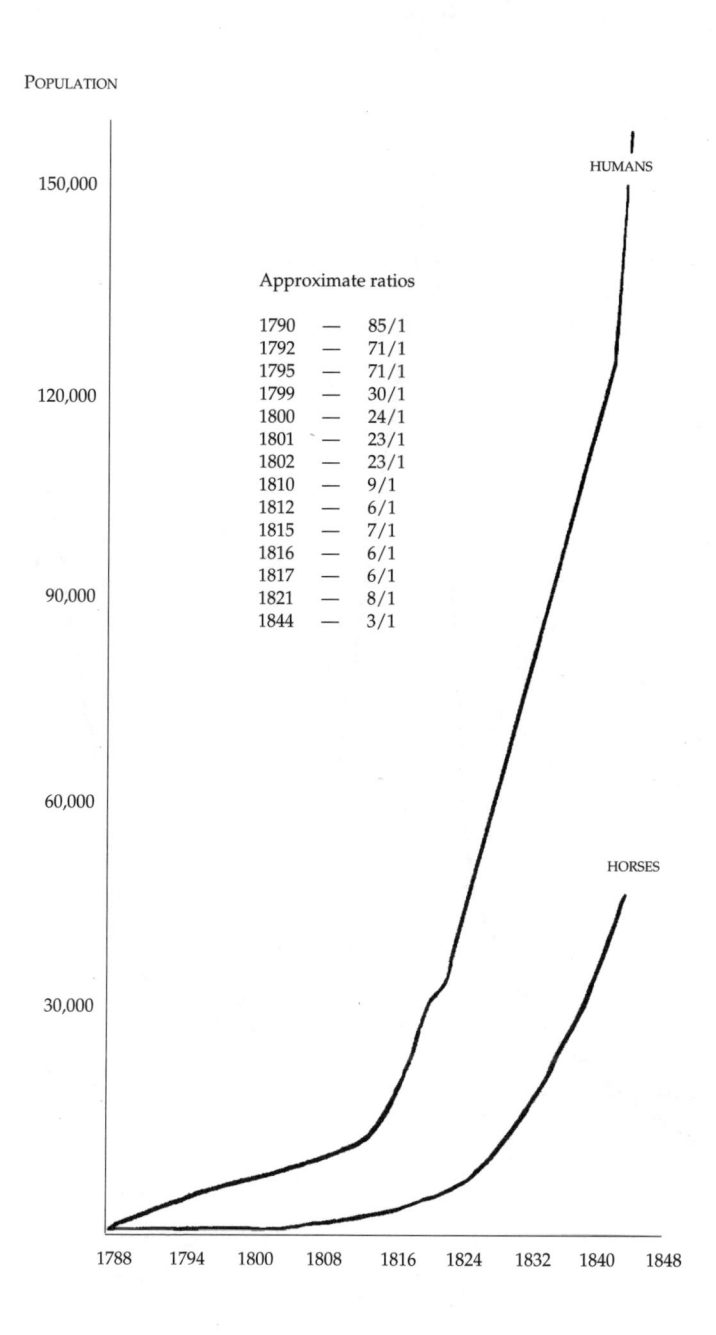

Year	Ratio
1790	85/1
1792	71/1
1795	71/1
1799	30/1
1800	24/1
1801	23/1
1802	23/1
1810	9/1
1812	6/1
1815	7/1
1816	6/1
1817	6/1
1821	8/1
1844	3/1

Figure 1: Horse–human population, 1788–1848
(Source HRN and HRA)

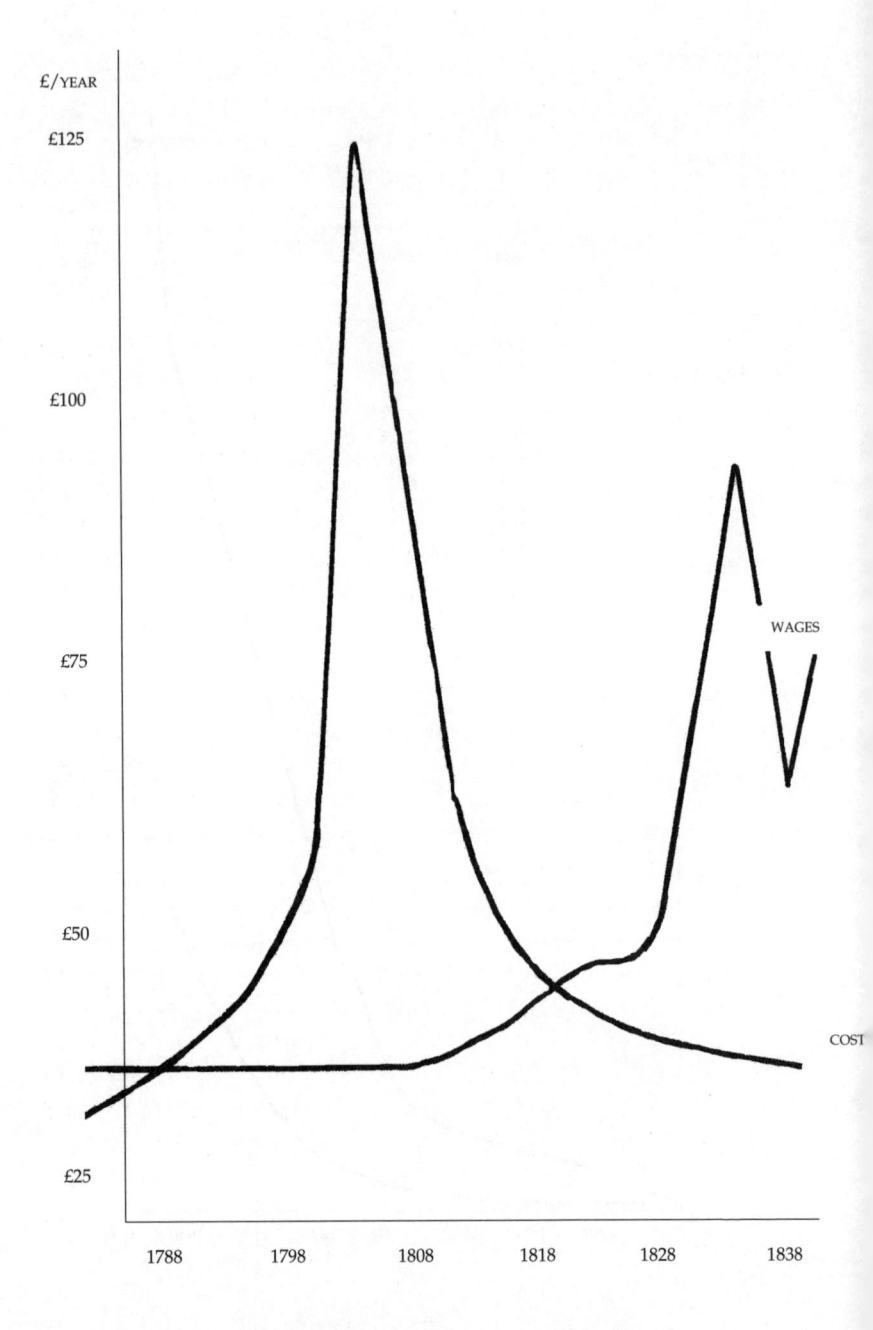

Figure 2: Wages and cost of horses, 1788–1848
(Source HRN and HRA)

about twenty-seven weeks. By 1844 there were 56,685 horses in New South Wales and prices had dropped to the point where, apart from bloodstock, importation was no longer economic.

Thereafter the growth of the horse herd was dramatic. In the period of pastoral expansion between 1860 and 1880, numbers leapt from 432,000 to 1,069,000, an increase of 247 per cent.[190] According to *Official Year Book* statistics, the Australian herd peaked at 2,527,149 in 1918.[191] It then declined continuously, and by 1959 totalled only 671,000.[192] The biggest reduction occurred during and shortly after World War II, when reliable heavy-duty vehicles and a vastly expanded road system were available.

In the early years following settlement, when transport was the major concern, the primary requirement of horses was considerable stamina. Much has been written on this subject and the commonly held perception is that, although Australian horses were renowned for their stamina for approximately the first half century, thereafter they declined markedly. The question bears examination, not only for historical accuracy but also to demonstrate how, if repeated often enough, ill-informed, subjective assertions can gain legitimacy and gradually be accepted by modern-day historians as authoritative statements.

According to accepted theory, initially stamina was not seen as a problem. Given the temperate climate, scarcity of worms and other parasites, and the superb natural grasslands of inland Australia, it is hardly surprising that early Australian-bred horses exceeded their parents in size and quality. Indeed, the comment was frequently made that these colonial horses were 'far superior to the original stock, both in strength and in beauty.'[193]

Ironically, such observations contradicted the prevalent British notion that England's soils and climate were so superior that both men and beasts inevitably degenerated once beyond Britain's shores. Thus it is hardly surprising that British migrants were amazed at the capacity of grass-fed Australian horses for repeated hard work. Emigrant mechanic Alexander Harris, for instance, wrote in 1827 that:

> most horses live on grass in this way for months together, and it is almost incredible what work they can perform. For instance, I have known a stockman ride his horse 60 or 70 miles a day, and with little abatement continue this for five or six days together; the horse all the time feeding only on grass and stabling in the bush.[194]

The highly experienced horseman, judge and commentator Edward M Curr provides a comprehensive and authoritative overview of the first four decades of growth and improvement. He states that fortuitously several of the earliest saddle-horse sires in Australia were particularly sound, strongly constituted and equable types, whose progeny demonstrated:

> a degree of stoutness and capacity for work quite unknown in England ... [and which were] remarkable for a vigorous health and freedom from sickness and disease.[195]

In this light he describes a 'rare stamp' of horse, whose performance surpassed even their fine looks. Most importantly, Curr corroborates what had impressed Alexander Harris so strongly: that working horses in Australia performed their feats of endurance while subsisting purely on the available grazing. 'The horse,' he states, 'gets nothing to eat but what he can pick up, which is often but a very scanty allowance.'[196]

Finally Curr points out that these horses were so well regarded that they were exported to India as cavalry mounts. He quotes one of the British Governors and his senior officers, who claimed that Australian horses, because of their size, endurance and weight-carrying capacity 'were preferred to all others at that time in India for that purpose.'[197]

In short, there appears little doubt that, within twenty-five to thirty years of settlement, horses of outstanding hardiness were being bred and raised in the colony. And yet Curr goes on to assert that within a short period the Australian saddle-horse began to deteriorate rapidly. Indeed, by the 1840s, when he first arrived in Melbourne, he claimed that stock coming from north of the Murray River were:

> generally poor, stunted, miserable wretches, the culls of that district, whose degeneracy weedy race-horses had begun, and early breeding and haphazard sires had completed ... The young stock had become leggy, girthless, boneless abortions, that had dwarfed down to ... fourteen hands [and] ... whose only value was their market price, for intrinsically they had none ... The little weeds now in use were unable to carry them, and the old settlers and Sydney natives looked back with regret to the old days when they rode those fast, up-standing, strong and fiery horses.[198]

Moreover, when the deterioration of the Australian saddle-horse became apparent in the 1850s, breeders attempted to reverse the trend. In most instances, however, they compounded the problem through lack of knowledge.

> More substance! big horses! began to be the cry. Away with these woolly-legged wretches! Let us have the tall bony horse of the old day! Breeders prepared to supply the demand. The weedy, delicate, washy, unsound, imported thoroughbred began to be put to the heavy cart mare. The produce of his cross was to be the sire for the future, and great results were expected. This sire showed an increase in size, his bone measured more than did that of his sire, he had some go, but still he failed to satisfy the public ... The broken kettle had been patched with wood and not iron.[199]

Indeed, by 1869 the horses being exported to India were such that one observer stated:

> It would be difficult to convey an idea of the unanimous manner in which the Australian horse is condemned as a coarse, awkward, ill-tempered, sluggish, and underbred brute ... [which displays on arrival] 1st. Reduced condition ... : 2nd. The large quantity of feed they need as compared with Oriental and Cape horses: 3rd. Their inability, without detriment, to stand exposure to the sun or endure long-continued hard work: 4th. Defects of temper, caused perhaps by injudicious handling as colts, but still existing to an extent to render many of them next to useless.[200]

Historian Ron Iddon argues that the situation had not improved by the turn of the century. He quotes the recollections of Chief Inspector of Stock Alex Bruce, who in 1901 also drew on the opinions of several other well-placed and presumably practically qualified commentators of the period.

> While we can with confidence give our saddle and cavalry horses of sixty years ago the high character we have done, we have to acknowledge with regret that the statement which we frequently hear, that 'Our saddle horses, from which the cavalry horses for India have for many years, and lately for South Africa, been drawn, have greatly deteriorated,' is borne out by the opinion of all those who are old enough to remember the horses of fifty or sixty years ago.[201]

Bruce and Curr agreed closely on the factors underlying the purported deterioration of the saddle-horse. For one thing, the demands of the gold rushes adversely affected the quality of horse husbandry. Many station hands departed for the diggings, leaving properties critically understaffed. As a result the routine castration of colts was frequently not carried out,[202] with consequent inbreeding and an accompanying deterioration of the quality of the herd. Furthermore, because of the rapidly increasing demand for draught animals, the strongest saddle-horse mares were now being mated to heavy horses, thus leaving the less robust mares to be mated with saddle-horse stallions.[203]

Another factor was the advent of thoroughbred racing. With the Australian emphasis on short distances and two-year-old events, the stallions available as station saddle-horse sires were often bred for speed rather than strength.[204] Paterson expresses this notion wonderfully in 'Old Pardon, the Son of Reprieve.'

> Three miles in three heats: Ah, my sonny
> The horses in those days were stout,
> They had to run well to win money;
> I don't see such horses about.
> Your six-furlong vermin that scamper
> Half a mile with their featherweight up;
> They wouldn't earn much of their damper
> In a race like the President's Cup.[205]

Furthermore, many of these stallions were the progeny of British thoroughbreds, which a prominent authority described at that time as becoming less and less sound.[206]

It must not be overlooked that the rapid growth of the colonies had led to a marked increase in the demand for horses of any description.[207] Thus, whereas earlier breeders could be discriminating in their selection of mares and stallions, those on large pastoral runs were frequently not now in a position to be particular, nor were they necessarily rewarded if they were.

In short, the evidence from apparently well qualified judges appears both to demonstrate their case and to justify the reiterative comments of modern-day writers such as Iddon:

> It is generally agreed ... that from around the middle of the last century, Australia's horses deteriorated.[208]

Yet a study of Australian exploration history and South Australian parliamentary papers produces contrary findings. In the first place much of the trouble associated with Australian horses in India can be attributed to the appalling conditions they endured en route.

> Mr Richardson, of the firm Taylor & Co., of Madras, who has upwards of twenty five years' experience of Australian horses, says that they are landed more dead than alive, though not in appearance; and only get fairly on their legs on an average in six months, with plenty of feed and proper looking after. He accounts for their arrival in this state from too great numbers being put into a ship, and sufficient attention not being paid to cleanliness. He has been on board several ships, and has seen some fifty or sixty horses standing with their feet buried in dung and filth; and the obnoxious effluvia arising therefrom made him feel sick for two days afterwards.[209]

Secondly, from the 1850s onwards explorers were attempting to traverse progressively more arid and difficult regions of inland Australia. According to Curr, Bruce, Iddon and others, increasingly these horses should have had difficulty coping with the environment and their workload. Yet the daily distances they covered showed a continuous improvement over those achieved by earlier explorers, who were working in easier country and were mounted on horses of purportedly greater endurance. Whereas, for example, fledgeling Western Australian bushman George Fletcher Moore was justifiably excited in 1837 at covering thirty-seven miles in a single day in the country immediately north of the Swan River,[210] the journals of Giles, the Forrest brothers and Stuart frequently mention journeys of forty miles per day through much more difficult environments, with forty-five to fifty mile days also being recorded from time to time.

Stuart and his associates made several references to the hardiness and reliability of the horses from the property of his sponsor, a Mr Chambers. A number of these animals were taken on several expeditions by Stuart and demonstrated a remarkable tolerance of heat, water stress and poor feed. According to the explorer, the horses to show out best under the worst conditions were those with an infusion of thoroughbred blood. These animals were more finely built and, while on physical appearances the coarser, more heavily built types would have been expected to cope better with severe stress, such was not his experience.[211]

Curr unwittingly provides an even more telling example of the hardiness of the Australian horse in the second half of the nineteenth century by citing an Arabian authority on the ability of the Arabian horse — a breed which Curr rated as far superior — to work for long periods of time without water. According to the Arab:

> Our horses could remain one or two days without drinking, once even passing three days without obtaining water.[212]

Yet Stuart, Giles, Eyre and, to a lesser extent, the Forrest brothers routinely pushed their mounts for up to three and a half days without water. Thus, although a considerable part of this improvement can be attributed to greatly improved horsemanship, whereby horses were carefully husbanded through arduous conditions,[213] the evidence shows that, despite the claims of apparently sound contemporary judges and their assertions regarding the increasing infusion of thoroughbred bloodlines, the quality of horses actually improved during this period.

Questions of endurance and resilience aside, with the rapid growth of the cattle industry horses increasingly had to be proficient at working cattle. For former drover Bruce Simpson, the characteristics of a good stock horse include that it is:

a fairly solid, well ribbed horse, with a good rein, withers, quarters and barrel, and fairly light legs ... [And] he's interested in cattle. If a newly broken colt is taken out around a mob of cattle and he looks at them, you know he has the makings of a horse. So it's the interest in the work. He has to be surefooted and fairly fast too ... He also has to have spirit. Station people have done a lot to destroy the Australian stock horse by getting rid of the ones that root [buck]. After three to four months' spell we expected a horse to root when they came in. If they didn't, there wasn't much guts in the horse.[214]

Interestingly enough, Simpson's assessment scarcely differs from North Queensland pioneer Edward Palmer's appraisal a hundred years earlier. Palmer touched on the necessary attributes, as well as the affection riders felt for such horses, in the poem 'My Old Stock Horse (Norman)', which he composed in 1894 in honour of his favourite horse.

> I have a friend — I've proved him so
> By many a task and token;
> I've ridden him long and found him true,
> Since first that he was broken.
>
> For twenty years we both have been
> In storm and sunny weather,
> And many a thousand miles we've seen,
> Just he and I together ...
>
> Across the Lynd and Gilbert's sands,
> And many a rocky river;
> Through trackless desert, forest lands,
> We've journeyed oft together.
>
> Then on the great grey plains so vast,

> Where the sun's rays dance and quiver,
> Through scorching heat and south-east blast,
> We've toiled on Flinders River.
>
> Through tangled scrubs and broken ground,
> We have often had to scramble;
> To wheel the cunning brumbies round,
> From where they love to ramble.
>
> Old Norman ne'er was known to fail,
> Or in the camp to falter,
> And just as sound to-day and hale,
> As when he first wore halter.
>
> Good horse, you have well earned your rest,
> Your mustering days are over;
> For all your time you'll have the best,
> And pass your life in clover ...[215]

Yet ironically, and somewhat illogically, horses such as Norman were almost invariably bred by chance. While the obvious approach would have been to breed from stallions and mares of good conformation, constitution and temperament, and run them on the most suitable country to ensure the foals were well fed, this was rarely done. In the first place, as with most aspects of horse husbandry, considerable disagreement existed over the ideal means of breeding and rearing. Curr, who observed and rode Arabian horses in Syria and other Eastern countries, was adamant that the best horses were bred in the hot and arid inland of Australia.

> The best of our horses in proportion to his figure, the most abstemious, stout, and sound, and the most neglected in his breeding, drinks of the waters that flow into Lake Alexandrina ... I feel no doubt ... that the time will come when the arid plains and burning

hill sides of Central Australia, will produce a horse that will be famous all over the world, and perhaps surpass even the Arab of the desert.[216]

Not surprisingly Paterson disagreed, arguing that:

Australian droughts occasionally play havoc with the young stock and the mares, and in bad years, when the horses have no milk, the whole year's crop of young stock are stunted ... [and these fillies then] produce small and stunted progeny ... The old Australian style of letting them shift for themselves and fight droughts, flies, and rabbits for a living [will] ... have to be abandoned.[217]

In fact Paterson claimed that, should Australian breeders continue to neglect their broodmares and foals, the Australian herd would inevitably diminish in size as had the South African Cape ponies. Both agreed, however, that shorter grasses were superior to long, rank, watery herbage (which usually grew in damp areas infested by worms and other parasites), that the worst horses produced were those bred along the Queensland coastal strip and other humid tropical areas, and that young horses reared on upland range country were the soundest and hardiest.[218]

A second major problem was finding suitable stallions. Indeed, given the explosive growth of the horse herd and the logistics of transporting large numbers of stallions, many of which had to be kept separate lest they injure or kill one another, it is hardly surprising that the early nineteenth century pastoral stallion was frequently just the best, or even the closest, male horse at hand.

The third and probably the most important factor affecting the breeding of good quality horses was the question of which were the most suitable breeds. In large part this was resolved by default. Once the racing industry became well established in southern Australia, the ready availability of thoroughbreds, plus the desire of many station managers to race horses at local meetings, meant almost all station sires in Australia thereafter were thoroughbreds or thoroughbred types. Certainly Indian Army Staff Veterinary Surgeon and remount agent W. Thacker found in 1874 that 'colts of unquestionable pedigrees and qualifications can be purchased for stud purposes in many parts of each Colony at reasonable prices ...'[219] Indeed several wealthy pastoralists believed so strongly in thoroughbreds that they imported stallions from England, specifically as station sires. Robert Christison of Lammermoor station in northern Queensland, for instance, was a 'great lover of horses and ran 500 ... which, in my time, were spoken of as "all thoroughbreds, bred from the best blood in England."'[220]

Curr lambasted this reliance on thoroughbred sires, arguing that, because of line-breeding for speed, these stallions were often temperamentally and constitutionally unsuitable. As he pointed out (apparently unaware of his contradiction), Australian horses exported to India were recognised as being 'exceedingly vicious.'[221] In short, Curr claimed the only redeeming feature of Australian stock horses was their underlying infusion of Arabian blood, arising from the Arabian parent stock of all thoroughbreds.[222]

Unarguably, 'blood' was not infallible and the upshot of purchasing unknown thoroughbred stallions from overseas or southern racing and breeding establishments was that:

> stock horses for cattle work are usually bred from sires with no special qualifications for that type of work other than speed. I never heard of a station stallion being chosen because it was good at everything required for cattle work, such as surefootedness and the ability to make a sharp turn naturally with its hindquarters ...
>
> The stallions used are chosen for appearance only, though sometimes their pedigrees are also taken into

> account ... But a racehorse gets no chance to show its ability at stock work and therefore goes out as a station sire, an unknown quantity at that work. It may or may not become the father of good stock horses.[223]

Thus it is hardly surprising that some proved quite unsuitable, siring vicious, dangerous and inherently bad horses. Reg Wilson's father, for instance, had considerable trouble breaking horses for E D White on a station near Charters Towers after he returned from World War I. These horses were by a thoroughbred or thoroughbred-cross stallion nicknamed the Snake and, according to Wilson senior, the progeny were 'snakes in temperament and agility, like their sire.'[224] Nevertheless, for all such criticisms, the thoroughbred type has served Australian cattlemen more than adequately. Indeed, as the widely experienced Simpson observes, 'I don't think the genuine Australian stock horse, the Waler type, could be bettered.'[225]

The other major breed of horse used in the pastoral industry has been the Arabian, or Arab as it is commonly known in Australia. Long recognised as extremely sound, hardy and equable, the breed also has remarkable weight-carrying and recuperative capacities.[226] The stallion Shahzada, for instance, successfully completed three 300-mile endurance races in India during the early 1920s before being exported to Australia as a twelve-year-old. He won two of these races, covering sixty miles per day for five days, and ran second in the other.[227] Not surprisingly, Curr claims that 'all Englishmen who have really tried the Arab at [distance riding] ... seem to come to one conclusion — that he has no equal.'[228]

Horses of Arabian blood were imported into Australia as early as 1803, when the stallion Old Hector was landed from India. A second Arabian stallion, Shark, arrived in 1804.[229] Their progeny were very popular, especially with the overlanders, who in the 1840s 'rode blood or half-bred Arab horses.'[230] By 1875 veterinary surgeon Thacker reported that:

he use of Arab sires was a favorite cross with many some years ago, and an animal extremely serviceable was the result ... and at the present time there is great demand for this class.[231]

The first professionally managed Arabian stud was established in South Australia by John Boucaut in 1891. The pedigrees of all horses bred on this property were recorded and, according to Coralie Gordon, for many years — indeed almost to the present day — as many Arabians as could be bred were sold as station sires to large inland cattle properties. In 1904, for instance, the Territory Government Resident reported that T H Pearce of Willeroo station, near Katherine, had purchased from:

> Sir J P Boucaut two desert Arab stallions, and he informs me that they are growing into nice horses — [they] are quiet and docile, yet full of fire. I think the cross he has decided on will result in a useful horse.[232]

Pearce's intention was to use these stallions to breed for the Indian remount market. This was a judicious choice, for in 1869 R D Ross had reported to the South Australian Government that the Indian Army preferred the true Arabian horse, bred for the most part in southern Persia, to all others.[233]

Yet paradoxically, despite their inherent hardiness and durability, Arab-type horses have not been found markedly superior under pastoral conditions. Furthermore, notwithstanding Gordon's claims, they have definitely not been largely accepted by the cattle industry. Certainly they have been well tried.

Among the factors underlying this apparent contradiction is the reality that, once the Australian herd had increased markedly, the number of Arabian horses imported was insufficient to have any noticeable effect.[234] Furthermore, most of the stock imported to Australia in the early to late 1800s were of poor quality. Captain Kent, for example, wrote to Governor

King in 1803 regretfully pointing out that 'it will be impossible to get anything like a good stallion of the Arabian breed under 200 pounds sterling.'[235] Given the previously noted wages of farm labourers at that time, such a horse would have cost approximately five years' wages or, in 1999 values, in excess of $125,000. In 1875 Thacker noted that, while the Arabian stallions imported possessed many excellent qualities, 'with very few exceptions, [they suffered from] the defective angle at their shoulder so common with them.' He believed their progeny were, because of this, 'bad hacks' and more likely to fall on rough going.[236]

While important in themselves, these factors were not the only reasons for the unpopularity of the breed. Later critics have attacked a variety of purported and actual faults, including that the anatomical configuration of their shoulders precludes them from galloping as fast as a thoroughbred.[237] This latter charge was addressed by Ion Idriess, who served with the Australian Light Horse in the Egyptian and Sinai campaigns of World War I. Idriess was contemptuous of the Turkish cavalry's Arabian chargers, frequently referring to them as 'little grey ponies,'[238] and recounts how Australian soldiers easily galloped down the sabre- and revolver-wielding Turks before clubbing them from the saddle.[239]

Surprisingly, given Curr and Thacker's insistence that Arab horses are outstandingly docile, the constant objection has concerned their temperament. Arabs often possess an extremely determined streak and when set on a particular course of action are not always easily dissuaded. Furthermore, even after gruelling hard work, they often will not settle down. Day after day they can cause trouble. Baker, former scrub-runner Robert Bradshaw, Ledger and Simpson all agree on this point.

The other major objection was that, although Arabs are highly intelligent and hardy animals, these traits do not necessarily produce outstanding cattle horses. Both Simpson and Ledger believe that overall they lack interest in working cattle and consequently do not learn to anticipate and place themselves correctly. As Ledger says, 'I can't ever remember seeing a decent face-of-the-camp horse being an arab.'[240] Consequently it is hardly surprising that Barker wrote in the early 1960s:

> At the present time there are some people who advocate the breeding of Arabs for station work. It was just the same sixty years ago, but still the proportion of Arabs on stations is about the same — only a fraction of the number of thoroughbreds in use.[241]

This contrasts strongly with the United States experience, where many nineteenth century working horses had a strong infusion of Arab blood, as well as Teutonic and Andalusian strains. These horses, which were often very small, were descended from the Spanish horses which Cortes landed in the Americas as early as 1519. The Spanish colonists attempted to upgrade the size and quality of their horses through stallions imported from the Spanish Royal Studs,[242] but little was achieved because of the huge areas and large numbers of horses involved, and the uncontrolled breeding of wild horses and Indians' horses. Accordingly, over three to four hundred years of indiscriminate breeding, the American horse became smaller and hardier. It also displayed a much wider variety of colours. Indeed, compared with the Australian herd, which has always had a strong infusion of thoroughbred blood and been relatively plain coloured,[243] the US herd has traditionally included appaloosas, duns, roans, piebalds, palominos and skewbalds.

Some regional differences of conformation also occurred, depending on the type of country and the infusion of differing breeds. Horses from the prairies and more fertile areas were well grown, whereas those of the arid south were frequently limited in stature by meagre feed. In fact ponies smaller than fourteen hands high were frequently used for cattle work in the south. This is particularly surprising, for Australian cattlemen have generally used a more substantial horse, one capable of coping with the rigours of heavy work in a harsh environment. Mora certainly appreciated that these were very small horses, and speaks of them as 'little rats.' But not all United States horses were as small. He also describes animals in the 'eight and nine-hundred pound classes' which, he admits, were necessary for 'real punishing work [where] ... you should have more weight to your stock horses to stand up well under heavy saddles and husky cowboys.'[244]

Attempts to improve the size of these small cow ponies through infusions of thoroughbred and Morgan blood[245] were not always successful, for big horses are generally less agile and surefooted. Furthermore, temperament was often sacrificed. Unlike Australian stock horses, US cow ponies have also to be broken to the roping and holding of cattle, which requires a docile and obedient horse. As Mora recalls:

> They started grading up, and some crosses took on the weight and size desired, and still kept, even gained, in speed. For flash reining, the kind 'that'll spin on a two-bit piece,' I think you've got to stop at a certain size and weight. For many years this type held down the range work, but in the last few decades the trend has been more towards the ... racehorse type ... It's hard to make ... a cow horse out of running stock or with too much hot blood. At least that's been our experience, and we've tried. It's true you need speed in a good cow horse, but you also should have a certain amount of phlegm and not too many hair-trigger nerves ... The majority turn out giddy headed, temperamental, excitable runners that aren't worth their keep when it comes to real cow work on the open range.[246]

Once these disadvantages became apparent, US cattlemen looked to other breeds, hoping that by judicious crossing they would incorporate the most desirable characteristics of several strains. To this end they even added a 'pinch of Percheron to add weight and to steady down some of the flighty hot-heads.'[247]

For over a century and a half Australian cattlemen have also experimented, trying to capture the better characteristics of a variety of breeds. Timor ponies were sometimes used for their hardiness and surefootedness. According to Wilson, his father's best mustering horse was a Timor cross and, although nearly all the wild descendants of these horses were shot out during the BTEC eradication scheme,[248] a few survivors can still be found on Cox Peninsula. Territorians Jim and Mick Fleming of Ooloo and Douglas stations also brought in big, solid remount stallions from the south, to breed army remounts. These were thoroughbred types, but bigger, heavier-boned and well sprung. Perhaps the most interesting infusion of bloodlines occurred at Dorisvale station in the Top End, where Wilson's uncle bred horses for the Pine Creek and Katherine police stations.

> They had the heavy horses and thoroughbreds and mixed them, and Timor ponies, which gave them hardiness ... so they were a combination of size, speed and stamina. Dorisvale horses were very good ...[249]

Clean-legged, light-draught breeds such as Suffolk punches and Percherons were also used to introduce more substance and bone into station horses. The heavier-bodied, coarser-shouldered progeny bred from such crosses, aptly named clumpers, were often used as packhorses, or wagonette or bronco horses. During the 1960s, for instance, the best bronco horses on Wave Hill station were half-Percherons sired by a Percheron stallion imported from England. Being a cooler blooded breed, they produced a more stolid horse, and:

> you could take them off the horsebreaker today, ride them out to the camp, put a collar on and bronco on them tomorrow. They were intelligent ... They weren't excitable, nor were they dopey.[250]

Furthermore, being heavier and more powerful than thoroughbred types, they coped better with the gruelling work.

This approach was broached as early as 1884, when Colonel Williams, the Director of the Indian Army remount operations, suggested upgrading the 'light weedy horses now being bred in Australia,' by crossing:

> the best of the light roomy mares you have with the best Suffolk Punch stallion you can get, and then put the thoroughbred horse to the filly produced, in this way getting a heavier strain into your stock ... [Furthermore,] if you study your own interests you will cull the fillies more than you do ... The fault of such a large preponderance of light stock being bred lies in the fact that so many of your mares are too light, and in many instances in the fact that there are a number of horses at the stud that have no business whatever there.[251]

When re-mated to a thoroughbred, clumper mares produced a reasonably fine-bodied saddle-horse that retained some of the robust qualities of its grandsire. The Peel River Land and Mineral Company, which bought the four Durack properties in the Kimberleys and owned Avon Downs and Headingly stations on the Barkly Tableland as well as several channel country and northern New South Wales stations, concentrated on the Percheron line. Under manager Dolf Schmidt, pure-bred Percheron stallions were used at the home property in Queensland to breed one-quarter Percheron stallions, which were distributed to the company's various stations.

Clydesdale stallions were also widely used, particularly in the Territory. Oenpelli Anglican mission at the East Alligator River used them from the 1920s onwards and, as pointed out earlier, Stan Bischoff bred one-quarter Clydesdale crosses for scrub-running in southern Queensland. For horses of their size, bulk and shoulder conformation, they produced comfortable and free-flowing riding horses.

United States quarter horses were also introduced to Australia from the late 1940s. Being about fifteen hands high, of solid build and with outstanding ability at turning, stopping and accelerating, this breed appeared to have much to offer. More importantly, their temperament was superb. In short, Barker believed that, 'as horsemen are not improving, it seems that here is an opportunity to breed a type of horse more suitable to present-day riders.'[252] Yet, despite their claimed advantages, quarter horses have not become predominant. Although well represented in Queensland, they have not been commonly adopted in arid regions such as the Pilbara, the Kimberleys and central Australia.

Finally, many properties also had a resident jack-donkey,[253] which was mated with station mares to produce mules. Mules are hardy and efficient foragers, have good weight-carrying capacity, are immune to many poisonous plants, including *Crotelaria,* and have naturally hard black hooves which do not need shoeing. They were used extensively in the Pilbara, Kimberleys and western regions of the Territory, but rarely in Queensland and the Barkly Tableland. Their use in these western areas may have been related to the harshness of the terrain, on which spinifex and scrub were often the only available forage.

Leichhardt, who was the first explorer to use mules, chose them primarily for their carrying capacity. On one occasion his mule Don Pedro escaped into the scrub heavily laden with pack-saddle and bags weighing 250 pounds, and roamed free for nearly a day. Yet when caught, to Leichhardt's admiration the beast 'trotted with the greatest of ease into the stockyard.'[254] Similarly, in his memoirs George Bates, who founded the major Western Australian firm of Bates Saddlery, writes that for part of World War I he was engaged in 'making [leather] shell carriers used for carrying 18 pound shells — four on each side of a mule.'[255] As each shell weighed approximately twenty-five pounds, the mules were carrying loads of two hundred pounds.[256]

Pat Underwood frequently employed mules on Inverway station, east of Halls Creek, especially towards the end of the dry season when temperatures were very high, the remaining feed poor in quality and the horses debilitated and poor in condition. In a tribute to their versatility, he used them for broncoing, packing, pulling wagonettes, mustering and cutting out. Retired drovers and stockmen testify to them being good, comfortable walkers. Other advantages were that standard saddles fitted them and they frequently displayed considerable intelligence. Edgar, for instance, had a small chestnut pack-mule on Wave Hill which:

> knew every dinner camp on the place. All you had to do was point the mule in the direction of the dinner camp ... and he'd go there and pull up at the same tree every time ... He was unreal, that little stumpy old bastard.[257]

Unfortunately, mules could buck ferociously. For part of the war Bates was stationed on Salisbury Plains in England, where his company took charge of a shipment of unbroken mules. These were to be broken to harness and pack-saddles, but:

> although we had some good riders in our unit, there was one outlaw who no-one could master. He was eventually disowned and turned loose on the plains.[258]

Furthermore, they could also be very difficult to catch. Wilson recalls one instance in the 1930s when his father was mustering at Burnside and using some freshly broken pack-mules. At lunchtime all the mules, bar one which was notoriously hard to catch, were unsaddled and tied in the shade. At this point Jack Noble:

the perspiration pouring from him, [said] 'Which pack bag has the rum. I want a nip badly.' My father pointed to the mule still dodging the would-be catchers and said, 'In there.' Jack said, 'Oh Arthur! I can't stand that — where's the rifle!' My father [then] said, 'No it's not. I'll get you a nip.'[259]

Horsemanship

Most Australians, certainly those who take motorised transport for granted, have no understanding of the limitations of horses. Continuous heavy exertion, especially when coupled with inadequate feed and severe conditions, soon drains a horse's vitality to the extent that it cannot cope with its workload. A horse in this condition is said to be 'flattened', and has to be turned out in a paddock and spelled until it has recuperated. This can take several months, depending on the available feed. Should circumstances dictate that the horse continue to be worked, it will gradually deteriorate to the point where it is incapable of any work and may even slowly decline and die. Accordingly, stockmen in northern Australia had to develop not only the unique skills required to manage their often poorly handled and extremely dangerous horses with safety and facility, but also the expertise necessary to keep them fit and sound for extended periods under arduous conditions. These skills, which are a combination of experience, intuition and empathy, are known collectively as horsemanship.[260]

In large part, early Australian horsemen had little expertise to draw from. Distance riding was almost unknown in Britain, and British horses were rarely ever taxed in the Australian manner.[261] Indeed, in Henry Handel Richardson's *The Fortunes of Richard Mahony*, Mahony is taken to task by Jopson, the local liveryman, for bringing his horses home 'winded and h'all of a sweat.'

> 'The 'ills is steep 'ereabouts, surr, and cruel 'aard on the 'arses. An 'tis naat the furst time neether. If you'll excuse me sayin' so, sur, them 'oove seen it do tell as 'ow you be raather a flash 'and with the reins.'
>
> 'Well, upon my word, Jopson, this is something new! I drive for show?... I overwork a horse? Why, my man, where I come from, it used to be dinned into me on all sides that I was far too easy with them.'
>
> 'Ca'an't say, surr, I'm sure.' ...
>
> 'But I can!' gave back Mahony, with warmth. 'I had two of my own there, let me tell you, and no beasts were ever better treated or cared for. They certainly hadn't to be walked up every slope for fear they'd lose their wind. They took their honest share of the day's work ... '
>
> 'Aye, surr, ahl very well, I dessay, for such a place — Australy, as I unnerstand,' answered Jopson unmoved. 'But 'twouldn't do 'ere, surr, in England. This's a civilized country.'[262]

Journalist, raconteur and bon vivant Nat Gould, writing of his experiences in Australia before the turn of the century, corroborates Richardson's belief, pointing out that:

> buggy-horses [here] are driven at a pace that would astonish anyone unaccustomed to the rapid mode of progression universally adopted ... There is not the

care bestowed upon horses in Australia there ought to be, and many of them are sadly over-driven.[263]

Nevertheless, Australian horsemen learnt to care for their horses remarkably quickly, especially considering the country they were often working in. Certainly by the 1850s they were proficient at assessing their stock's condition and husbanding it over long distances.

In large part the horsemanship essential in the arid regions of inland Australia was initially developed and refined by explorers such as Eyre, Stuart, the Gregory brothers and Giles. W P Auld, who in 1861 accompanied Stuart on his successful fifth attempt to cross Australia, recalls that:

> Stuart was a wonderfully good judge of the condition of his horses. When out exploring in splendid grassy country which looked as if water must be near, he without any apparent reason would give orders to return and we found we had just as much as we could do to get back without losing any of our horses or of our men.[264]

Perhaps the best example of the proficiency of Australian horsemen during this era occurred during Augustus Gregory's 1855–56 North Australian expedition. On this occasion, the inexperienced German botanist Ferdinand von Mueller lacked the practical skills necessary to husband his horse's strength. Thus, although he was lighter than all the other more experienced members of the party, he knocked up four horses in seven months whereas not one of the others' mounts became incapacitated during this period.[265]

To some extent this expertise was maintained by drovers, who had to carefully husband their horses for months on end over hundreds, even thousands of miles. Unfortunately it appears that in some regards horsemanship underwent a decline from the 1870s onwards. Mustering cattle was markedly different from exploring and settling new country. Day after day of hard galloping was often required and stockmen were well supplied with horses. Accordingly, they could afford to be less concerned with husbanding them and it is not surprising that Australian horses in the Boer War and the early years of World War I stood up little better than any other nation's. Quite simply, despite frequently being superb riders, Australian light-horsemen generally lacked the skills necessary to maintain their horses over extremely arduous and extended campaigns. Indeed, during the Boer War Paterson wrote 'many of [our] horses are dying, weakened by hard work, under-feeding, colds and sore backs,' and then went on to say:

> The Australians have not shown up as the best horse masters by any means ... This question of sore-backed horses and lame horses is one of the military questions of the present day, and we Australians, for all our horsiness, are just as far off solving it as the Imperial Tommies.[266]

Nevertheless, it is informative to examine the extensive range of factors explorers, stockmen and light-horsemen learnt to consider and the skills they developed in their attempts to keep their horses sound and fit for extended periods of work. The first and most important consideration was careful selection. Horsemen had to learn how to evaluate an animal's soundness, constitution and temperament. Soundness is not overly difficult to determine, for visual evidence is usually available of current tendon, ligament, knee, pastern or hoof problems, or a predisposition towards them. The explorer Charles Sturt also found from experience that horses must be absolutely free from blemishes on the back and wither, for these were signs of former galling — that is, the flesh being rubbed raw by saddlery or harness. Such marks not only indicated a predisposition towards galling, but were also likely to gall again. Certainly by the early 1900s horsemen had become particularly acute at detecting warning signs. Paterson, for instance, mentions how it was:

> marvellous to see a buyer for the Indian market stand in an Australian stockyard and point out at a glance defects in wild, unbroken horses as they scamper past him in a cloud of dust.[267]

Intangibles such as temperament were a different matter, requiring considerable experience. Yet horsemen soon understood that the shape and cut of a horse's forehead and facial bones, and the softness of its eye, coupled with its general demeanour, are accurate indicators of temperament. By the 1860s they had also learnt a number of the indicators necessary to assess an animal's constitution. Perhaps more importantly, they understood that horses subjected to unremitting hard work also required a degree of psychological toughness. Leichhardt and fellow explorers Sturt and Stuart all agreed that 'it is in a great degree the spirit, which carries man and beast through difficulties.'[268]

Given the requirements of rest and recuperation, horsemen also had to learn how to calculate the demands of the task in hand and thus the number of horses required. This aspect was clearly understood as early as the 1860s. On his fifth attempt to cross the continent, Stuart took seventy-two horses. His decision to take so many meant that loads were kept light and the vital and extensive reconnoitring was shared across a greater number of mounts.

Stockmen also learnt to cover only short daily distances early in a journey while their stock were soft in condition, and gradually increase the workload. Racehorse trainers have understood this principle for some centuries. This was not always possible in stock camps, where the first day's work could call for many hours of hard riding, but horsemen were certainly aware of its desirability. Furthermore, in order to avoid flattening their mounts, horses were generally changed at lunchtime. Riders also trotted a lot, for trotting is the most economical gait for the horse, although it means the rider has to work harder. Curiously, US cowboys never rose to the trot. For some reason, no matter how rough the horse was at trotting:

> you just rode it anyhow, glued right there and not the slightest rising in the stirrups ... That was only made for dudes and Englishmen — it was taboo on the range.[269]

In his 1889 novel *Robbery Under Arms*, Rolf Boldrewood touches on how inexperienced or uncaring riders can quickly flatten a soft horse. In this case Captain Starlight's half-caste Aboriginal assistant Warrigal arrives well after nightfall to fetch the Marsden brothers. After a brief rest, they saddle their horses and leave before dawn. Because of long-standing animosity, Warrigal, hoping to show up the Marsdens' horses, sets a very strong pace in front on his hard and wiry horse, Bilbah. Dick Marsden, knowing their horses are not fit enough to continue at this speed for many miles, remonstrates with him.

> 'Now, look here, Warrigal,' I said. 'You know why you are doing this, and so do I. Our horses are not up to galloping fifty or sixty miles on end just off a spell and with no work for months. If you don't pull up and go our pace I'll knock you off your horse.'
>
> 'Oh, you're riled!' he said, looking as impudent as he dared, but slackening all the same. [I would have] pulled up before if I knowed your horses were getting baked. Thought they were up to anything, same as you and Jim.'
>
> 'So they are. You'll find that one of these days. If there's work ahead you ought to have sense enough not to knock smoke out of fresh horses before we begin.'[270]

Preventing galling was also vitally important, as the pain it causes can lead to bucking, savagery and debilitation. Indeed, on one occasion during Surveyor-General John Oxley's 1817 and 1818 explorations of the Lachlan and Macquarie Rivers, after his pain-ridden horses were unsaddled, they all lay down:

> like dogs about us ... The backs of the greater part of them were, not withstanding every care, dreadfully galled, so that they could, when first saddled, scarcely stand under their burdens ...[271]

Probably the first recorded account of galling in inland Australia can be found in Assistant Surveyor-General George Evans' record of his 1815 crossing of the Blue Mountains and exploration of the hinterlands. On this occasion his horses became badly galled on their withers and he was forced to give them days off, slowing his rate of progress.

Galling can also permanently incapacitate or kill animals, should the condition known as a fistula become established. A fistula arises from scalding caused by profuse sweating under the harness in the vicinity of the wither or shoulder, or an abrasion from ill-fitting saddlery or harness, which ulcerates and becomes pustulous. Due to the vertical musculature of much of the shoulder, gravity progressively carries the pustulous ulcer downwards. As it slides between muscle sheaths, a membranous lining often develops, forming a discrete tract which heals very poorly. In most instances sections of poorly vascularised bone and cartilage along the wither and shoulder become infected, and extensive surgical removal of infected tissue is often the only treatment option.[272]

As a result of this and other experiences, horsemen developed and refined a number of techniques for reducing the occurrence and severity of galls. The foremost of these lay in the careful selection of stock. Well-conformed horses in good condition rarely gall if normal precautions are taken. By contrast, pot-bellied, narrow-chested, weedy or immature animals will often gall, for their conformation allows the saddle to slip forward and the girth to ride up into their armpit.

Having well-lined and correctly stuffed saddles and collars was also crucial in the prevention of galling. Evans' problem with galling occurred because the straw packing of his pack-saddles was not lined with serge. To counteract the roughness of the straw lining, Evans and his men used their woollen blankets under the saddles.[273] Similarly, saddlery with stuffing which has become hard through being compressed by continuous use is more likely to cause galling, so re-stuffing is important. As early as his 1848 explorations of the Shark Bay and Murchison River regions of Western Australia, Augustus Gregory set time aside to re-stuff unsatisfactory saddles.

Later, in the Top End, preventing scalding from excessive sweating during the wet season became particularly important, and stockmen and station saddlers selected the straw used for stuffing with care. On Burnside station, what was known as 'bladey grass' was used. This is a tall grass which grows on the margins of swamps and watercourses. Being quite fibrous, it allowed better air circulation, which helped keep the packing drier and cooler.

An adequate saddle-blanket was also found to be important in preventing galling. Augustus Gregory, knowing his horses would have to cover thousands of miles during his 1855–56 North Australian expedition, stipulated 'heavy saddle-blankets.' A variety of materials have been tested over the years, and heavy woollen or felt blankets have proved highly suitable and durable. The choice of wool is not surprising, for, although woollen blankets do not absorb sweat, neither do they gall horses to the same extent as cotton, once wet. Yet no matter how suitable the material, galling can still occur if the blankets are caked with an abrasive mixture of dried sweat and dirt, and horsemen rapidly learnt to pay particular attention to the cleanliness of their saddle-blankets. In many instances they scrubbed them after each day's work and hung them to dry overnight. Similarly, leather girths were kept clean and well oiled to ensure they were soft.

Early explorers and bushmen, who often worked in regions where replacement horses and bullocks were not readily available, also quickly became aware of the necessity of keeping their stock free of dried sweat and dirt. Dried sweat can lead to galling, for it abrades the skin when harness is placed over it again, and needs to be carefully removed. Thus part of the everyday care of working horses entails cleaning those areas under the saddle or harness that have sweated badly. Leichhardt took especial care to keep his horses' backs carefully washed, but, because he had no spare mounts, most of them eventually became badly galled. If water is not available, brushing is equally satisfactory. Augustus Gregory appears to have been the first explorer to appreciate and practise this, for he included twelve brushes in the equipment for his North Australian expedition.[274]

Not overworking stock while their skin was still soft was also important. Eyre, who was particularly aware of the problem, commented in the early 1840s:

> I was anxious to move quietly on at first, that nothing might be done in a hurry ... Nothing is more common than to get the withers of horses wrung, or their shoulders and backs galled at the commencement of a journey, and nothing more difficult than to effect a cure of this mischief whilst the animals are in use. By the precautions which I adopted, I succeeded in preventing this, for the present.[275]

Ideally, should galling occur, the animal was rested. Continuing to work an animal that has galled, especially in hot weather when it is sweating copiously, can cause the wound to worsen and enlarge until the animal eventually either loses condition or bucks when harnessed. In this regard Aboriginal stockmen were often held to be very lax.[276] Usually, however:

> We never rode a horse if it became scalded. We

> washed their backs every day and if there was any type of a blemish you'd give him a spell for a few days. You'd be lucky to use one horse twice a week because you probably had four to five horses to choose from.[277]

An oil- or grease-based remedy, such as Bickmore's Salve, Flint's Oil or Pottie's Hopple Chafe, was then applied to keep the wound supple as it dried, to prevent scar tissue forming and to keep flies away.

Packhorses, which rarely exceeded a walk, faced particular problems. It was critically important that they were not overburdened, and much of this expertise was gained at the cost of terrible suffering by early explorers' horses. Oxley's cruel and dangerous overloading of his packhorses during his 1817 and 1818 explorations exemplifies this learning phase. Once his expedition left the Lachlan in 1817, the stores were divided up and sufficient rations taken for the estimated duration of the exploration in front of them. At the time the daily military ration was one and a half pounds of flour and approximately three-quarters of a pound of salted pork. Using these amounts as a base, the gross requirements for the party could have been calculated and their weight established. It would have then been possible to determine the required number of packhorses. Such a procedure was not followed, however, and an arbitrary number of horses seem to have been taken. Indeed, the evidence suggests that there were fewer than half the number required for the task, each horse being forced to carry over three hundred pounds. As a result they were all exhausted after only four days' travel from the Lachlan and within eighteen days three had either died or been shot because of exhaustion-induced paralysis of the hindquarters. Such exhaustion is inconceivable under normal circumstances.

On his second expedition, Oxley made his horses carry even heavier loads and literally broke their hearts. He was well aware of the inhumane demands he was making of his animals.

The horses were, however, very heavily laden, carrying at least 350 pounds each; a weight which I was fearful the description of country we had to pass over would render still more burthensome. We had, however, relinquished everything that was not indispensable, and the saddle-horses were equally laden with the others.[278]

The results were predictable and, given the saturated, marshy country they were attempting to traverse, worse than on the first expedition. After only two days the horses were exhausted and 'failed much during the day, and several of them were severely wrung with their burthens ...'[279] A fortnight later, in one terribly marshy section, the exhausted horses fell continually. Oxley later noted that they covered only six miles for the day and 'for the last half mile, the horses were not on their legs for twenty yards altogether.'[280]

By contrast, as early as 1845 the widely experienced Gregory brothers were calculating the number of horses required on the basis of the weight of the equipment and provisions that were to be taken. By the late 1850s they understood quite clearly what a horse could be expected to carry and what equipment was required for an extended period of time. Augustus Gregory knew, for example, that as supplies were consumed and the gross weight declined, the horse's carrying capacity was in most cases also declining. Accordingly he never entertained the idea of overloading the horses in the earlier parts of the journey in anticipation that the gross weight would soon be reduced.

> A pack horse does not carry more than 150 pounds exclusive of packages and harness and even when the quantity is being reduced by the regular consumption of provisions, the casualties and reduction of strength from fatigue, reduce the means of transport to nearly an equal amount. Thus 3,300 pounds would require 22 horses for its conveyance, each carrying 150 pounds.[281]

Unfortunately, Oxley's ignorance of loading was compounded by cruelty. On the first expedition he mentions that his horses were carrying pork barrels and that on marshy country they were continually falling. Although such behaviour could simply indicate exhaustion, as pork barrels then weighed up to 320 pounds,[282] it is also evidence of an extremely unbalanced load. Load balancing is a critical factor in keeping packhorses in good health. Unbalanced loads lead to galling of withers and girths, because the girth straps have to be pulled up tighter than is desirable to ensure that the saddle and load do not slip. As the horses could not possibly have carried two barrels of pork in order to balance their loads, a single barrel was somehow strapped onto one side of the pack-saddle, leading to severe imbalance as well as extreme overloading. As a consequence the horses suffered from frequent falls and terrible galling, and the expedition was continually forced to make short daily stages. On occasions, Oxley even had to spell the horses for one or two days, thus unnecessarily wasting time. Clearly his animals were unfortunate guinea pigs in the early usage of packhorses in Australia.

The Gregory brothers, however, were particularly aware of the need to balance saddle-bags and husband their stock. As supplies and equipment were consumed, loads were continuously and precisely re-allocated with the aid of a spring balance. Furthermore, on his North-West Expedition Francis Gregory assigned the men to work in groups of two, so the heavy saddle-bags were loaded on both sides simultaneously. In this way the saddles and saddle-blankets were not dragged to one side during the loading process, thus reducing the possibility of saddle-blankets being creased and galling the horses. Even the minutiae of everyday care were conscientiously observed. During Augustus Gregory's North Australian Expedition each man was allocated a group of horses, and was responsible for their general care and the maintenance of their saddles and harness. This encouraged the men to take better care of the horses that were their specific responsibility.

With time, stockmen also learnt to stop and give their packhorses a short break every two hours or so. Not only did this give them a brief rest, but it also allowed them to urinate. Most stockmen unsaddled their packhorses at lunchtime, or at least removed the saddle-bags, in order to give them some rest. Alternatively, the day's journey was completed in one stage, in which case the horses were out grazing some hours earlier.

Although shoeing working horses came to be an important feature of Australian exploration and pastoral life, this was not always the case. Leichhardt, whose horses and bullocks frequently suffered appallingly from footsoreness, never appreciated the necessity of shoeing his stock. During his first expedition he took only one spare set of shoes for each animal, despite originally calculating that the expedition would last between five and seven months. He then threw the spares away at the Staaten River, in order to lighten the loads. Leichhardt later commented that his stock had frequently been badly footsore and that periodically he had to halt to rest them for a few days. In fact, after one particularly torrid day spent toiling over the hard basaltic outcrops which cover much of the bed of the Lynd River and its surrounds, another expedition member commented that 'we could have been tracked for miles over the Rocks by the blood which came from the poor Animal's [feet].'[283]

In Central Australia, Sturt found the intensely hot, dry sand badly affected his unshod horses' feet. Indeed, on one occasion their hooves became so dry 'that splinters flew from them at every step.'[284] Eventually, in November 1845, he was forced to leave behind a horse which was incapable of keeping up with the main party. It had worn down its feet so badly that the sensitive frog — a triangular rubber-like structure at the rear of the sole — had been torn from one hoof 'and you could see a large raw cavity up the heel.'[285] In fact, both Sturt's and Leichhardt's stock would have been walking on the soles and frogs of their hooves. Not surprisingly, these animals were often intensely footsore and as a result their rate of travel was reduced.

A reduction in mobility is not the only concern with

footsoreness, however, for badly footsore animals are less able and willing to graze. Francis Gregory demonstrated this quite clearly in his 1861 North-West Expedition. When returning to his depot after an arduous traverse of the southern Pilbara region, he had no shoes to re-shoe his horses. Accordingly, when he was forced to keep pushing them across the rugged ranges immediately east of Nickol Bay, 'some of [them] ... walked upon stones as they would over red-hot coals.'[286] The worst affected horse had already been left behind, 'the hoof being fairly worn through,'[287] and when Gregory returned through the area eighteen days later, he commented:

> [I had] hoped to find him improved by the rest; but, on approaching the spot, the presence of crows and a wild dog gave indications of a different fate; we found him partly devoured within a few yards of where we left him, inflammation of the feet having most probably produced mortification.[288]

Thus it is hardly surprising that shoeing became commonplace. The highly proficient bushman and explorer Augustus Gregory, for instance, always kept a watchful eye on his horses' hooves during the North Australian Expedition. On rocky ground between the Roper and Albert rivers, he reported that 'the day seldom passes without having to replace the shoes of several of the horses.'[289] Fortunately he was well prepared, having brought six hundred sets with him from Sydney and manufactured more at a depot he had established on the Victoria River.[290] Similarly, Reg Wilson recalls that even at the depths of the Depression his father ensured that the front feet of his best horses, which were doing the bulk of the hard work on the rocky limestone and basalt country, were shod.

The principle underlying shoeing is not complex. Briefly, the exterior of a horse's foot consists of a very hard, dry, insensitive hoof wall, a more sensitive and flexible sole, and the frog. These structures are of differing hardness and elasticity, not only to

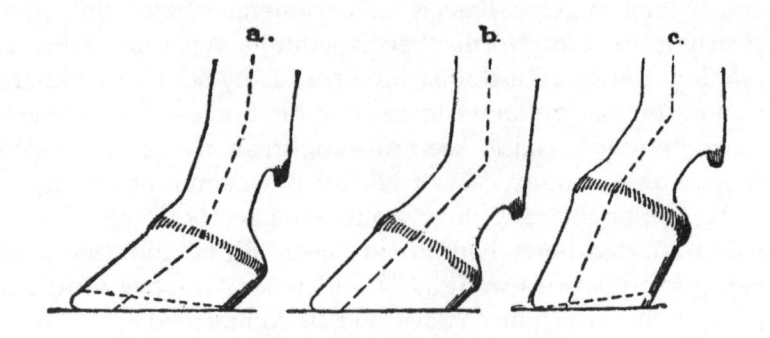

—a, Side view of foot with the foot axis broken backward as a result of too long a toe. The amount of horn to be removed from the toe in order to straighten the foot axis is denoted by a dotted line; *b,* side view of a properly balanced foot, with a straight foot axis of desirable slant; *c,* side view of stumpy foot with foot axis broken forward, as a result of overgrowth of the quarters. The amount of horn to be removed in order to straighten the foot axis is shown by a dotted line.

Plate 34: Examples of correct and incorrect hoof axes.

allow for expansion and contraction of the hoof as part of the weight-carrying and shock-absorption system of the lower leg, but also to provide protection.[291] A line of cells within the coronary band, at the top of the hoof wall, is the growth site for the outer, horny hoof. On average, the hoof wall grows about one-quarter of an inch per month, and ideally it should be of sufficient size to lift the more sensitive sole and frog clear of the ground by about one-quarter of an inch. On softer ground, such as the plains country of much of central and western Queensland and inland New South Wales, this rate of growth is sufficient to replace the hoof being abraded. On hard or rocky ground, however, the rate of wear often exceeds the rate of growth, and the outer wall gradually becomes shorter until the horse is carrying its weight on the sole and frog. These soft structures are not intended to directly carry weight and, should the horse tread on a stone, the sole takes almost all the weight pressing against the sensitive inner sole causing internal bruising. Stone bruises can make a horse sore and unwilling to walk, and may also turn septic, forming an intensely painful abscess which will lame the horse completely for some time. Accordingly, steel shoes are nailed to the bottom of the insensitive outer wall, thus lifting the sole and frog about three-eighths of an inch away from the ground.

Different techniques of shoeing were soon developed for horses doing different work. Theoretically, a line drawn up the front of the hoof wall and continued along the pastern should be straight. If the toe or heel is too long the line will not be straight, as shown in Plate 34, and the configuration so formed places an abnormal strain on the ligaments and tendons of the lower leg. Farriers thus trim the hoof to obtain the correct conformation. Notwithstanding this ideal, stockmen learnt to trim the toe of the front feet back hard and scarcely touch the heels, when surefootedness and rapid changes of direction were required. By this method the horse stands more upright, thus allowing it to roll onto the toe and break over more easily and quickly while galloping. Polo horses are shod on the same principle. Horses working on basalt country in Queensland were often shod with oversized shoes which had the heels turned inwards until they almost met, in order to protect the sensitive frog.

While well-handled horses, which have had their feet lifted and tapped as week-old foals and regularly trimmed thereafter, are rarely any problem to shoe, such was not always the case with poorly handled station horses. Frequently they were terrified at having their front legs lifted, and reared or pulled away. Restraining the horse by holding an ear helped, but the assistant ran the risk of being struck. Lifting the hind legs was almost invariably considerably more difficult. Even if a stockman was able to lift a hind leg from the ground, he then had to extend it backwards in order to rasp the foot and nail on the shoe. In doing so he was very badly placed, for the horse

Plate 35: Dave Ledger holding down a stubborn horse's head in the channel country after it had been thrown, early 1960s.

could leap forward slightly and then kick with both hind feet at maximum strength.

Accordingly, stockmen devised a number of protective techniques. Usually a collar-rope or strap was placed on the horse and a long rope noosed tightly around the pastern of the hind foot to be shod. This was then looped through the collar-rope and the leg pulled well forward and tied up short, leaving the horse balancing on three legs and more amenable to standing still. In this position, however, the farrier could not see the bottom of the hoof, and both rasping and nailing were done more roughly. A neater job could be made if the foot was not pulled so far forward, for the bottom of the hoof was more accessible, though this approach was more difficult for the farrier as he had further to bend.

With time and constant shoeing, most horses become more

amenable, and the task becomes progressively easier. Some horses never improve, however, and require throwing every time they are shod. A number of simple yet highly effective techniques were developed in Australia. Retired drover Bruce Simpson describes how a long rope was tied in a loop around the horse's neck and the ends taken around each hind pastern and back through the neck-rope. These two ends were then shortened simultaneously, pulling the hind legs forwards until the horse was forced to sit on its hindquarters. It was then rolled over on its side and the leg-ropes tied to the neck-rope. In this position the farrier could not be kicked, but neither could he shoe the horse as well. This method was best used on horses which would tolerate ropes near their hind legs.[292]

The horse shown in Plate 35 was thrown by this technique in the channel country in western Queensland during the early 1960s, when it would not accept the saddle-bags after being saddled. As can be seen, the horse was first hobbled and neck-roped, then its back legs were pulled up to the neck-rope. The saddle-bags are in the background. Similarly, the horse in Plate 36 has been thrown and both front legs and one hind leg tied together with a hobble strap, and its head is being held to prevent it attempting to regain its feet. As a result the stockman doing the shoeing has a relatively safe task, notwithstanding that the hoof he is nailing is still free to kick.

Barker outlines an alternative technique he used on a horse he had purchased cheaply because it was known to be a rogue at shoeing. After failing to get ropes anywhere near the wildly kicking animal's hind legs, Barker and his offsider were advised by their elderly partner to sit down and have a pannikin of tea. Afterwards, the older man put hobbles on the horse's front legs. He then tied a long rope to the hobble chain and, making a wide berth around the animal's hindquarters, secured the other end to a nearby tree. Using a long branch, he then gave the recalcitrant animal a vigorous cut across the hindquarters. Leaping forward, the horse quickly reached the end of the rope

Plate 36: East Arnhem Land coastal Aborigines shoeing a horse, c. 1930s.

and cartwheeled forward onto its shoulder. Unable to rise, it thrashed with its hind legs until one of them became caught over the hobble chain. In this position it could neither rise nor free its hind leg, and the other leg was then roped and also pulled up short. Now being unable to kick or strike, the horse was shod on all feet.[293]

Footsoreness amongst bullocks also had to be taken into consideration. On his Port Essington expedition, Leichhardt's older, heavier bullocks became more footsore than the younger, lighter ones. As station owners usually selected old, heavy bullocks — known as 'fats' — for market, it is hardly surprising that a proportion of them became badly footsore, especially when traversing areas covered with sharp-edged pieces of shale. Frequently a drover would notice a bullock becoming increasingly lame and lagging at the rear of the mob. Eventually the beast's progress would become so slow and painful that shoeing was necessary. Only the outside branch of a bullock's cloven hoof was shod in the process known as cueing. While special shoes were available (see Plate 37), drovers and teamsters traditionally used worn-out horse shoes. These were snapped in half at the front and the appropriate section nailed to the bullock's left or right hoof. Small shoeing nails were used, a bullock's hoof wall not being as thick as that of a horse. In describing this process, Barker claims that, because bullocks never became quiet enough to shoe unrestrained, a crush or pen was necessary.

Cueing pens used to be a common sight, rigged at convenient places. They consisted of a strong post and rail 'V'. The bullock was pulled into the 'V' with a rope attached to its horns and tied to the centre post. Behind the bullock was a strong, short rail rigged only

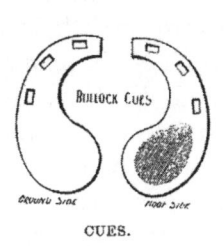

CUES.

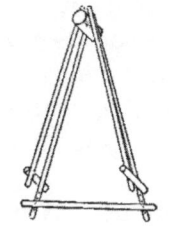

BIRD'S EYE VIEW OF PEN.

Fit the cue as tightly as possible, three nails, arranged as illustrated, being enough. Always put the nose nail in first and drive it well home before putting in the other two. A pen for tying up bullocks is shown.

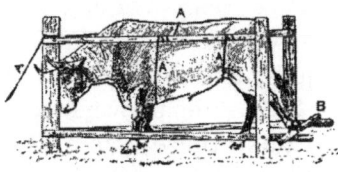

BULLOCK IN PEN.

Plate 37: Bullock cues and cueing frame.

two feet from the ground. A bullock's hind foot was pulled back to that rail by a rope and fastened with the front of the foot resting over the rail. [See Plate 37.] Then the cues were nailed on, no rasping or fitting, and were done quickly. For cueing the front feet a rope on one foot went over the bullock's shoulder to a man standing on the opposite side. He'd pull until the knee bent and the hoof was at a convenient height for shoeing, then tie the rope to the other front leg on which the bullock was standing.[294]

Generally, however, drovers were a little less particular. Schultz found that by the time a bullock's feet were badly sore it was also tired from travelling and amenable to submitting to cueing. On the first occasion he decided to cue a bullock, he had two bronco mules in his team, so looped a rope around the bullock's hind leg and had the mule pull it out backwards. Another stockman pulled a front leg out forwards and they rolled the beast on its side, when a third man held it down. Peter Ross had the most casual approach of all, merely getting an Aboriginal stockman to throw the bullock. Whatever the method of restraint, the cues were nailed on the uppermost hoof, then the beast was rolled over on its other side and the opposite hoof cued. Within four to seven days, once it had overcome the internal bruising and general inflammation, the bullock would be seen striding out freely.

Another major aspect of maintaining all working stock lay in maximising their opportunities to graze. Stock working all day and relying purely on local feed require several hours of grazing each day to satisfy their requirements. This placed early stockmen in an awkward position. If the animals were left free to graze overnight in unlimited bush there was a strong likelihood they would wander and be lost, whereas tying them up overnight left insufficient time for grazing.

One of the earliest recorded instances of stock wandering can be found in Evans' journals of exploration. He did not fetter his

Plate 38: Hobbles, also known as hopples.

horses on either his first or second expedition and on one occasion they wandered overnight. They had been tied together but 'not secured in a proper manner,' presumably by tying them to a tree. Consequently most of the following day was spent searching for the animals, which were eventually found with the ropes fortuitously wrapped around a tree. Later in the same year, during Governor Macquarie's tour of the new settlement at Bathurst and its surrounds, the party's cart-horse broke free one night and wandered twelve miles back along their tracks. This mishap entailed a walk of twenty-four miles for the hapless groom, following which Macquarie solicitously commented, 'This will be a lesson to Joseph during the rest of our journey, to be more careful in tethering his horses.'[295] Bullocks were found to be even more troublesome than horses, and a sure source of time-wasting and delay until they settled into their new station and ceased wandering.

Within a decade, however, effective methods of restraining stock had been developed by the new generation of bushmen who were pushing further into inland New South Wales. A book outlining bush life in the early settlement, entitled *Settlers and Convicts: or Recollections of Sixteen Years Labour in the Australian Backwoods*, written by 'an emmigrant mechanic', believed by Manning Clark to be Alexander Harris, is a useful source of information on the techniques used in the bush at that time. It outlines how by the late 1820s pastoral stockmen, who generally worked and camped well away from stockyards and fenced horse-paddocks, gained the advantages of both overnight grazing and restraint by hobbling their horses. A hobble consists of a chain about a foot in length with a swivel in the middle and a leather strap at either end. Plate 38 indicates how the straps are buckled around the pastern of each front leg. Although the short length of chain prevents the horse from walking normally, it can still graze satisfactorily by shuffling forward. Indeed, hobbled horses can cover up to eight miles a night and even canter. In fact, on Francis Gregory's 1861 exploration of the north-west of Western Australia, six of his hobbled horses actually galloped into the camp from where they had been grazing. Admittedly they had a strong incentive, for hard behind were eighteen armed Aborigines who had already attempted to spear them.

Hobbling appears to be derived from the British technique of spencilling, by which horses, bullocks and milking cows were fettered with a leather strap or short-noosed rope. The technique was employed in Britain as early as 1610[296] and was used to fetter front or hind legs, or the nearside or offside legs. The latter configuration, shown in Plate 39, is called a side-line in Australia.

With hobbling and sidelining, finding stock in the morning was less of a problem. The task was made easier when stockmen learnt to set out after their horses very early in the morning, before they had re-commenced grazing and wandered

Plate 39: A side-line.

further away. By this stage they had also become proficient at tracking.

> When turned out for the night the horses seldom stray far; they are hungry and tired, and like to make the most of their time on any patch of good grass they come to. But if a young horse does happen to walk off, he is easily tracked by the experienced eye of the bushman, and hobbled as he is, is easily overtaken.[297]

Curiously, although the hobbling of stock at night soon became standard practice, bells were slow to come into use. In forest country bells save a considerable amount of time, especially when several horses' tracks have criss-crossed as they graze throughout the night, or where rocky or hard ground makes it difficult to track them. The impoverished Leichhardt was able to afford to bell only one horse on his first expedition. Unfortunately, during August 1845 the clapper broke when the party was west of the future settlement of Burketown in the Gulf of Carpentaria, and the bell was largely ineffective thereafter. In memory of this incident he named the site Camp of the Broken Bell. By contrast, as expeditions from the 1850s onwards were equipped with necessarily larger numbers of horses, considerably more bells were taken. Augustus Gregory, who was acutely aware of their value, took twenty bells with him on his North Australian Expedition.[298]

In arid regions the provision of water was also a vital aspect of horsemanship. Severe dehydration was a common problem and on many occasions horses knocked up and even died. Giles, for instance, was beaten back from the fringe of the Gibson Desert in 1873 and wrote movingly of the surviving horses' determination to reach water.

> The pack-horses now presented a demoralised and disorganised rout, travelling in a long single file, for it was quite impossible to keep the tail [of the mob] up with the leaders ... The real leader was an old black mare, blear-eyed from fly-wounds, forever dropping tears of salt rheum, fat, large, strong, having carried her 180 lbs at starting and now desperately thirsty and determined, knowing to an inch where water was; on she went ...[299]

Both Eyre and Sturt learnt that extremely thirsty stock will not graze and will thus lose condition rapidly. In situations where survival was finely balanced, inducing stock to graze could be the difference between life and death. Both men would walk their stock until they were tired, then give them a limited amount of water from a supply they carried on their drays. The stock would then graze and rest for some hours before they were again given water, reharnessed and set under way. By using this technique Sturt was able to cover the last 112 miles of waterless country on his return journey from Central Australia in 1846 and reach the safety of the Darling River. Similarly, on his North-

West Expedition, Francis Gregory found that his debilitated horses coped better if he matched his travelling and camping schedule to available water supplies. He therefore camped beside water during the hot part of the day, set off again in the late afternoon, camped for the night without water further along the track, then got under way again very early when the morning was still cool. As pointed out in the chapter 'Droving', this technique was later generally adopted by drovers, who rested their mobs alongside water in the middle of the day.

During this period of exploration it was also found that experience allowed both horses and cattle to better tolerate severe water stress. On several occasions Stuart observed that his older horses, which had suffered from deprivation on previous occasions, coped much better than the less experienced animals. Thus, although he felt sorry for his horses when they were suffering badly from thirst early on the second expedition, he saw value in the experience, noting 'I expect they will endure it better next time; they now know what it is to be without.'[300]

Nevertheless, there were limits, and stockmen in inland Australia needed to be able to assess how much further their animals could go without water. Although it is commonly accepted that it is not desirable for working horses to go without water for more than thirty hours in hot weather, Eyre was concerned but not perturbed when he made his horses go for up to four days without water. He was confident of being able to nurse them through such stages and on one occasion recorded the crossing of:

> upwards of 100 miles of desert country, during the last three days, in which the horses had got nothing either to eat or drink. It is painful in the extreme, to be obliged to subject them to such hardship, but alas, in such country, what else can be done?[301]

Experience also showed that when water was finally found, horses had to be held back and watered carefully and frequently, being only allowed small amounts at a time. Should they be given unlimited access whilst still severely dehydrated, they were likely to suffer badly from colic.

Horsemen throughout Australia also learnt to deal with ailments such as wounds and colic, although in most instances treatment was quite unsophisticated. Indeed, in 1885 Alfred Giles of Springvale station, Katherine, reported that the manager of Wangalary station:

> in common with others, suffered severe losses amongst his horse stock from gravel disease. This is caused by an insufficient supply of salt. The horses lick imaginary salt patches, and swallow sand, stone and gravel, which ultimately kills them. No cure has yet been found, but I have heard that kerosine and soapsuds have been used [as a harsh purgative] with good effect.[302]

Eighty years later Underwood's pharmaceutical supplies still consisted of only 'a bit of kerosene, corned beef fat and Stockholm tar, for everything.'[303] Generally, stake-wounds were poulticed with sugar and soap and proud flesh with bluestone (copper sulfate), stone bruises were cut out and the horse shod, eye infections from blight or grass seeds were treated with sugar and buffalo fly wounds to the corner of the eye with iodoform and castor oil.

In the northern pastoral regions horsemen faced additional problems, not the least of which was the complete lack of professional advice on lameness. In most instances they would never have seen how to flex pasterns and knees, nor how to apply pressure to tendons, suspensory ligaments, sesamoid bones and hooves in a systematic attempt to isolate the source of lameness, despite there almost always being:

> a lame horse in a stock camp. He's either put his foot in a hole galloping after cattle at night, and there's always antnests, stakes or holes in the ground.[304]

Furthermore, because most stations had an abundance of horses, there was little need to do so: 'If they went lame you never heard words like shin-soreness. You'd just tip them out [in the paddock for a spell].'[305] Bowed tendons were treated with petrol or methylated spirits as a crude working blister.[306]

Kimberley walkabout disease, the first recorded case of which occurred on Alexander Forrest's 1878 De Grey to Port Darwin expedition, also defeated most northern cattlemen. On 20 June 1879 Forrest reported:

> One of the horses last night showed symptoms of madness and wandered away ... He was tracked, but was presumed to have stumbled in a creek and been drowned.[307]

According to former boss drover Matt Savage, horses died in their hundreds from this disease and many theories were advanced as to its cause. Savage quite perceptively noted, however, that the highest incidence occurred close to rivers, and that by keeping the animals away from river country and onto areas of black soil they would be spared. Research by the Western Australian Department of Agriculture in the 1970s verified that the disease stems from eating the native legumes *Crotalaria retusa* and *Crotalaria crispata*. Ingestion of these plants causes liver failure, which results in a build-up of ammonia in the bloodstream and consequent intoxication of the central nervous system.[308] The affected horses become disorientated and walk compulsively until their eventual death from debilitation, or from starvation resulting from the associated loss of vision. The Department of Agriculture corroborated Savage's observation, noting that *Crotalaria* grows almost exclusively on sandy soils and that by running horses on harder country the problem can be avoided.

The veterinary condition anhydrosis, known in outback Australia as puffs, was also of serious concern to northern cattlemen. Affected horses have partial to complete inability to sweat, due either to clogging of the sweat glands from abnormal sweating or a reduced sensitivity of sweat gland adrenoreceptors from constant overstimulation. Strenuous exertion is not the sole cause, for relatively sedentary pregnant mares are also affected. High temperatures and humidity are implicated, and thus the highest incidence occurs in the humid coastal areas of the Kimberleys, Top End, Gulf and Cape regions. As a result of these factors, body heat cannot be lost by evaporation and, in an attempt to dissipate heat, affected horses breathe very heavily with flared nostrils for up to an hour after exercise — thus the term 'puffs'. Affected horses have also been known to immerse themselves in water for some time in an effort to reduce their body temperature. Exercise tolerance decreases according to the inability to sweat and badly affected animals can scarcely be ridden. This complaint is also common among racehorses exported to South-East Asia. They are treated with electrolyte drenches and ceiling fans in their stables, or transferred to the cooler and less humid highlands. Clearly these options were not available to northern cattlemen, and affected horses, along with those suffering from walkabout disease, had to be replaced.

Finally, being able to identify poisonous plants was important in almost all areas of Australia, and the tropics were no exception, as the earliest explorers and pioneers found to their cost. Leichhardt, for instance, lost a horse at the junction of the Lynd and Mitchell rivers in Queensland in 1845 and attributed its death to snakebite. Later he assumed that the three of his horses found dead on a sandbar projecting into the Roper River had been drowned by crocodiles. Yet it is almost certain that all four were poisoned. Strong evidence that the first horse met its death in this way can be found in the journal of the Jardine brothers' expedition in the same area twenty years later. They lost over ten of their best horses, and described the effects of the poison as it ran its course on their stock. More importantly, they verified Leichhardt's earlier mention of:

the loss of Murphy's pony on the Lynd, which was found on the sands, 'with its body blown up and bleeding at the nostrils.' Similar symptoms showed themselves in the case of the horses of this expedition, proving pretty clearly that the deaths were caused by some noxious plant.[309]

By contrast, the West Australian Gregory brothers were vigilant about poison, for large areas of the south-west of Western Australia support native plants which contain lethal levels of sodium fluoroacetate. Problems with such plants date from the mid 1830s, when several groups of settlers were working their way south from Perth to the Williams River district and one party lost eight out of ten bullocks immediately inland from the Darling Scarp.[310] Although the Gregorys did not lose stock during their early Western Australian expeditions, they commented several times on the presence or absence of the *Gastrolobium* species. Thus, when a horse was lost on the North Australian Expedition, von Mueller examined its gut in order to ascertain which plants it had eaten, for Augustus Gregory wished to avoid losing additional stock in this way.

Saddlery and Harness

Like all other facets of pastoral technology existing at the time settlers first crossed the Blue Mountains, saddlery and harness had to be rapidly modified. Fortunately the experimental ideas of stockmen and horsebreakers were readily incorporated into general practice, as saddlers and harness-makers began operating in the Colony as early as 1819.[311] Indeed, for more than 150 years these men not only played a vital role in the development and working of stations, but were also of considerable importance to local and state economies. Collectively they used immense amounts of leather. Each large stock saddle, for example, used one side of leather, or half a hide. From the 1920s through to the 1950s, heavy and light leather was produced by about twenty major tanneries throughout Australia. Apart from one in Western Australia and two in South Australia, these were located in the eastern states. Saddlers and harness-makers also purchased very considerable quantities of associated equipment. The chain manufacturing firm Pitt Waddel, for instance, made almost all the hobble, spring cart, plough and dray chains required in Australia, while other companies produced the swivels for hobble chains. Saddle-tree, collar-hame and saddlers' thread manufacturers also played important roles.

Most saddlery construction was carried out by large firms in the major cities and business was often brisk. George Bates, for example, writes that in Western Australia during the 1920s:

> the Nor-West was being developed, [and making] donkey and camel harness was keeping harness and collar makers fully employed. Sir James Mitchell, the Premier at that time, was mad on land development. Soldier settlement and group settlement was taking place in the South-West. Migrants and returned soldiers were being put on the land ... [and] every settler was given a set of harness ...[312]

Accordingly, there were about eight major saddlery shops in Perth and two to three hundred people working in the industry. Of these, only fifty were making saddles, the remainder being involved in the manufacture of rugs, boots, bridles and so forth, as well as working for specialist harness and collar shops.

During the 1930s Depression, however, the saddlery industry was very badly affected. Bates speaks of sweatshop conditions, where tradesmen had to produce six hand-made saddles each week and 'many a young saddler got the sack for failing to finish his six ...' Costs and quality were slashed in an attempt to counter Sydney-made saddles, which 'looked nice, with clean leather and a nice finish ... [but which] were shockingly made, [with the] cheapest of trees, and thrown together.'[313]

By the late 1940s business had declined to the extent that only Bates and Arundels were still manufacturing saddles in Perth. Yet less than a decade later, when the industry was beginning to recover from the decline in horse numbers and World War II, Syd Hill Saddlery in Brisbane were employing 120 staff and producing three to four thousand stock saddles annually, as well as making race and show saddles and carrying out repairs.

Saddlery shops also played an extremely important role in country towns. In 1896, as a result of closer settlement and a large cattle herd, Queensland had in excess of 370 saddlery businesses for a population of 449,174.[314] Country saddlers did mainly repair work, being too busy to make saddles. Many larger stations also had resident saddlers to repair bridles, re-line saddles and make the huge number of hobbles required each year.

Where there was insufficient work for a resident saddler, travelling saddlers were often called in. These were not common in Western Australia until after World War II, due to the huge distances and poor roads.[315] They were, however, a common feature of pastoral life in the eastern states. Furphy mentions the fictional 'Joseph Pawsome, saddler and evangelist' travelling in an emblazoned wagonette throughout the Riverina in the 1890s and Jim Hill, former manager of Syd Hill Saddlery, recalls being told by his grandfather of:

> a travelling saddler out Cunnamulla way. He had a big barrow and pushed it from station to station. He'd write to Brisbane ordering leather and equipment to be delivered at a certain rail siding by a certain date, and would collect it and push it in his barrow to the next station. That was in the 1880s.[316]

Although Australian saddlery was modified considerably to suit outback conditions, the underlying design and construction principles remained remarkably constant for almost 180 years. Being based on English patterns, all Australian saddles up to the early 1970s were built around a wooden frame, called a 'tree', which was joined together by glue, screws and small steel straps (Plate 40).

Nearly all trees produced in Australia in the twentieth century have been manufactured by a small number of

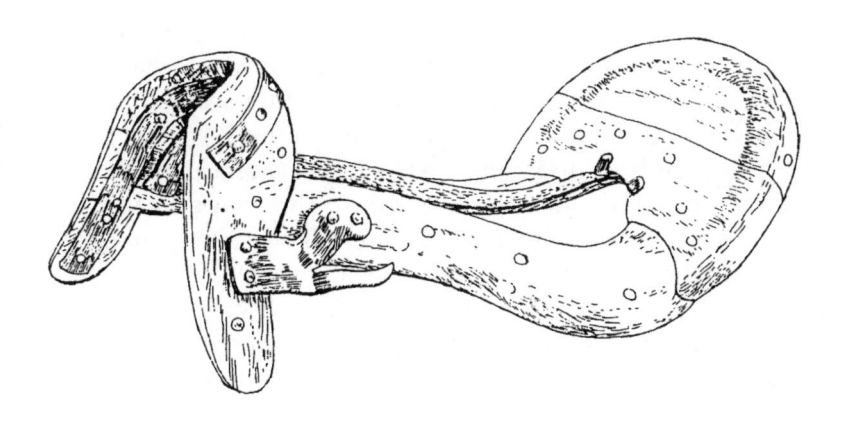

Plate 40: A saddle tree.

specialist city firms. They are cut to stock patterns according to differing requirements and sold bare to saddlers. From the 1930s onwards there were only two tree-making firms. Andrew Gabb, which operated from Canterbury in Sydney, became Andrew Gabb and Bagnall in the early 1940s, while the Bright Brothers worked from Brunswick in Melbourne. The latter firm made trees only for show and race saddles, with all stock saddle trees coming from Sydney.

The trees were covered in a criss-cross pattern by strips of webbing attached by screws and tacks. The webbing formed the basis of the seat, and also the attachment point for the girth straps. The heavy leather seat and flaps were then stitched together and attached to the bare tree, after which the saddle was stuffed.

Traditionally, horsehair was used for stuffing, although, as Don Bates explains, 'there's a myth about horse hair.' Because of cost the initial and main stuffing was actually done:

with cow hair, called doe hair ... The horse hair was only used as a facing. All horse hair stuffing would have been too springy and the saddle would have moved on the horse's back. A new saddle was stuffed with doe hair, then used for some months. It was then brought back and hair-faced with horse and pig hair, the doe hair having settled into the shape of the horse's back. Horse hair was inserted via tufting holes and the hole was quilted to make it invisible ...

Horse hair varied in price and quality. The good saddles had the better horse hair, which was five to ten times more expensive ... Most 'horse hair' was [actually] pig hair. All good horse hair was used for paint brushes and was too expensive for saddles. Scrap horse hair was mixed with the trimmings from pig carcases.

A Melbourne firm made the horse hair for saddles. It had to be washed, boiled and then curled into a rope to get it to spring. In its natural state horse hair has no spring and will very quickly compress flat and provide no padding for the horse. It was cut into the appropriate lengths as required and teased out.

We used horse hair mattresses, made in the 19th century for hospitals [to prevent bedsores]. When they were sold, as spring mattresses were introduced, we bought them, washed and treated them, and used the hair in saddles.[317]

By contrast, station saddlers usually had a cheap and ready source of horse hair. Cattleman Frank Dean of Strathdarr station recalls how brumby shooters in the Longreach region during the 1930s were paid £1 a head and also took the mane and tail. This and other hair pulled (thinned) from the manes and tails of newly broken horses was collected and carefully picked clean by Aboriginal women before being washed in a copper with warm water and soap for several hours. After drying, it was twisted into approximately two-foot lengths with a carpenter's brace and then tied off. These lengths were then placed in clean water in the copper and heated, but definitely not boiled, for several hours. They were subsequently dried over a fence and stored in the saddler's room until required. During the wet season, when all saddlery and harness were overhauled, these lengths of washed hair were untied and teased out. By this process, the required 'spring' was attained.

Yet, despite the overall uniformity of design and construction, major external modifications were essential during the first half of the nineteenth century, for the two types of British saddles then available were inadequate. Cavalry saddles, although large and heavy, were only lightly padded and notorious for causing galling. Curr, in describing the arrival of a troop of mounted police at his Port Phillip property in the 1840s, commented on their 'awkward saddles, and sore backs.'[318] Nothing had changed sixty years later when Banjo Paterson, writing his dispatches as a Boer War correspondent, mentioned how almost all Colonel Knight's horses were critically affected by sore backs, because they had been 'sent from Australia with saddles I wouldn't put on a mule.'[319] George Bates confirmed Paterson's opinion of these saddles.

> I think the worst designed saddles that were made in Australia was the military saddles ... made for the South African war in 1899. It is said, when Lord Kitchener saw them in Africa, he put them into a heap and burnt them.[320]

The relatively flat, light British hunting saddle was also unsatisfactory, being uncomfortable for both horse and rider over extended periods of work. Of course exceptions could be found, depending on the skill of the saddler. As well, saddles with a broken tree were less rigid and often softer to ride in.[321] Perhaps the most humorous account of this is provided by Furphy when he has Tom Collins accompany Jack the Shellback to the boundary gate. As they saddle their horses, Collins stares in amazement at Jack's saddle, which is:

> a second-hand English saddle, of more than ordinary capacity. The barrow-load of firewood which had once formed the tree was all in splinters, so that you could fold the saddle in any direction; and the panel had from time to time been subjected to so much amateur repairing that, when Jack mounted, he looked like a hen in a nest, so surrounded he was with exuding tufts of wool, raw horse-hair, emu's feathers, and the frayed edges of half a dozen piles of old blanket, of various colours. But when he said it was the softest saddle on the station, though it would be nothing the worse for a bit of an overhaul, I was bound to admit that the statement and the reservation were equally reasonable.[322]

The major objection to both the army and the hunting saddle was, however, that they provided little assistance when dealing with the generally unhandled and often dangerous horses encountered on pastoral properties. On being mounted, these animals frequently bucked violently, and riders using either type of saddle were often soon dislodged (Plate 41). Indeed, one of the riders in the World War I Army Remount Unit in Egypt described 'those slippery army saddles' as 'patent self-emptiers.'[323]

Bucking appears to have been relatively unknown in England, for several British commentators of the period mention it. The Reverend William Haygarth, who spent eight years travelling through inland eastern Australia in the 1850s and 1860s, best expresses their considerable surprise.

> Australian horses have a vicious habit known as 'buck-jumping' or as it is more familiarly called, 'bucking.' This trick, in its aggravated form, is

Plate 41: A sporting rider being summarily dislodged from a low, flat English hunting saddle.

peculiar to the colts bred in the colony and in Van Diemen's Land, and is decidedly the most expeditious way that could be devised for emptying a saddle ...[324]

Furthermore, these early saddles often shifted forward over a horse's wither as it bucked. In this situation a rider finds it very difficult to get his feet free of the stirrups and, as the saddle moves further onto the horse's neck, he may fall in front of its shoulder. He is then liable to be trampled or kicked viciously as the panic-stricken horse bucks itself clear. Clearly, some form of restraint was required to prevent saddles moving and to reduce the likelihood of riders being propelled forward or backward as their mounts bucked and reared wildly.

The first modificatory response to these shortcomings was to use a crupper[325] and improvised kneepads when riding rough horses. As Curr notes in his treatise *Pure Saddle Horses, and How to Breed Them in Australia*, riders learnt from hard-earned experience to pass a crupper strap:

> between the saddle and saddle-cloth until it obtrudes before the pommel,[326] it is then passed two or three times around a stick about 20 inches long, and as thick as the wrist, called the 'kid', and there buckled, [and the] kid is strapped to two iron staples, with which the saddle is provided, behind the pommel. This kid comes across the horseman about six inches above the knees, and adds very much to the security of his seat …[327]

While such rudimentary kneepads would have been extremely uncomfortable for the rider, they were preferable to being thrown on hard ground. Thereafter kneepads became increasingly important.

Possibly the earliest mention of leather kneepads can be found in the letter written at Moreton Bay in February 1859 by British migrant and pioneer grazier Ernest Henry. Henry, who settled on Mount McConnell station at the junction of the Burdekin and Suttor Rivers and later in the Mount Isa region, asked his father to forward a few specifically modified saddles. He had quickly learnt the necessity of having his saddle flaps cut further forward. More importantly, he requested his kneepads be not only twice as large but also contain three times as much stuffing.

Initially, leather kneepads were soft swellings fitted underneath the saddle flap in the manner of English hunting and hacking saddles. In this configuration they provided little assistance to a rider slipping forward, so were quickly transferred to the outside of the flap, in much the same position. Accordingly these early kneepads were known as knee-riders,

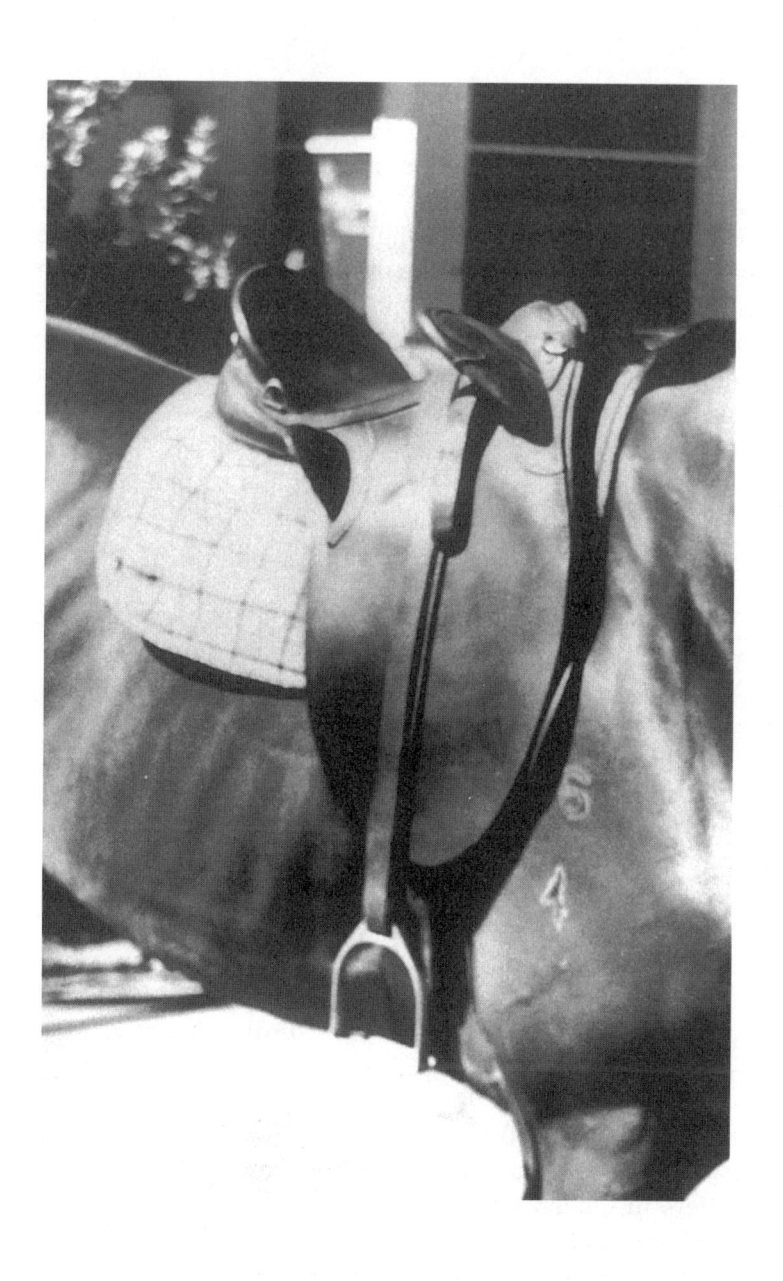

Plate 42: A Wieneke horsebreaker's saddle with raked kneepads.

for a rider's knee fitted into the pad. Obviously, in this low-set position they provided little assistance and thereafter kneepads were set increasingly higher on the saddle, so the rider's thighs pressed against them.[328]

Despite being fixed increasingly higher on the saddle, these early kneepads were constructed of a leather outer casing stuffed with doe hide which meant they bent if a rider was thrown against them by a bucking horse, thus largely negating any advantage. Kneepads changed dramatically in the 1920s, when saddlers realised the advantages of solid, all-leather construction. In his memoirs, George Bates writes:

> I think I can lay claim to have developed the solid leather knee pad ... I was able to use up tons of scrap leather from the harness and saddle departments. I was criticised at first, as it took a little time to perfect the pattern. It turned out to be much quicker, and [made] a firmer and more solid pad. It kept its shape better and I was given five shillings over the award rates.[329]

Jim Hill observes, however, that Bates' claim is arguable and that he may only have been the first in Western Australia. Indeed, Bates does admit that:

> a saddler named Smith opened shop here. He came from Queensland. He helped me to alter the style of saddles, especially kneepads. I studied his saddles and learnt much.[330]

Kneepads gradually grew in length, with most averaging between three and a half and five inches. In the early 1920s saddlers in Queensland developed a long backward-raking style of kneepad for horsebreakers (Plate 42). These were up to six and a half inches in length and, although a bucking horse drove a rider's thighs under them and sat him more firmly into the seat, they inhibited his freedom to move in accordance with the horse's motion as it rapidly changed from bucking to leaping to rearing. Furthermore, pads of this nature hindered a rider in trying to get free of a horse that had overbalanced while rearing and was toppling over backwards. In fact two riders were killed in this manner in Western Australia during the early 1960s, and as a result Bates ceased production of that style of pad. Saddlers have subsequently refined the shape of kneepads, which now roll slightly forward at the top, thus reducing bruising to the rider's thighs.

Although it took Australian saddlers almost a hundred years to incorporate high kneepads, it is surprising that the United States cowboy, with centuries of experience to draw from, did not develop a counterpart to kneepads until around the turn of the century. Unarguably they realised the necessity for such assistance. Former cowboy Jo Mora mentions how, prior to their introduction, when assigned a horse he believed would cause trouble, he:

> never hesitated for a second to take every advantage I could. Generally I tied a folded and rolled slicker or mackinaw back of the slick forks for a bucking roll.[331]

In the early 1900s Mora and a companion, who were at a tavern in Holbrook, Arizona, celebrating the end of a trail drive, were talked into walking down to the nearby railway siding to look at a radically different saddle. Although initially sceptical, Mora soon realised that this innovative saddle was the long-awaited replacement for his mackinaw 'bucking roll.' While different in detail and structure from Australian kneepads — US 'kneepads' are actually a swelling on either side of the fork of the tree — their effect is the same: they reduce the likelihood of a rider being thrown forward by a bucking horse. Mora notes that these protuberances, which they termed 'swelled forks' as distinct from the earlier 'slick forks', have been constructed as wide as twenty-two inches (Plate 43).[332]

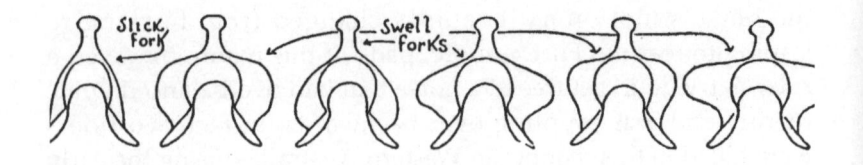

Plate 43: Front sections of a variety of US saddle trees, showing the roping horns and 'swelled forks'

Further modifications which led to the development of the Australian stock saddle included increasing the height of both the back and front of the seat — the pommel and cantle — to reduce the likelihood of a rider slipping as a horse reared or bucked (Plate 44). Different materials were also incorporated, including rough hides. According to George Bates, rough hides were first used during the early 1930s when Dale's Tannery of Botany experimented with tanning the rough inside of hides because they could not procure sufficient hides without scratches on the outside. Saddlers then constructed a few saddles using rough hides. In fact rough hides were being used at least eleven years earlier than Bates estimates. In 1919 the *Northern Territory Times* ran an advertisement for Edward Butlers' saddles, in which the seat was stated to be constructed of 'Rough Cut Bullock Hide.' Presumably some riders believed these saddles gave greater security and made the difference between staying on and a heavy fall, for orders increased steadily. A similar situation existed in the US, where saddles with the rough side out were believed to be 'fine bronc saddles.'[333] Thus it is understandable that slippery new saddles made from traditional smooth leather were treated with a liberal coat of tacky dubbin or home-made saddle dressing.

At much the same time a number of older stockmen, many of

Plate 44: An Australian stock saddle with high kneepads, cantle, pommel and flank girth strap.

whom suffered from haemorrhoids because of an inadequate diet and the physical strain of their work, ordered saddles with pigskin seats. According to Hill, pigskin breathes and was less irritating for them under hot and sweaty conditions.

Moreover, in the 1940s Bates, who was then working for Perth saddlers Arundels, received a request from renowned Kimberley cattleman Tom Quilty to design a saddle that would prevent his Aboriginal stockmen giving their horses sore backs. Over time, numerous references have been made to this problem throughout Australia. Generally, they were based on the tendency of many Aboriginal stockmen to ride loose and with long stirrups, so throwing their weight onto the back of the saddle. Accordingly, Bates made a saddle he called the Quilty, which was 'designed to keep them well up into the saddle …'[334] The fact that such a saddle would then place his Aboriginal stockmen at a disadvantage when riding a bucking horse appears to have been immaterial to Quilty. In fact the pastoral industry had a tradition of differentiating between white and black stockmen on the basis of their saddlery. As early as 1919, for example, Edward Butlers of Brisbane were advertising their 'Black Boy's Saddles' for £6/15/-. As Butlers' other saddles were retailing for £12/12/- at this time, the quality of these cheaper saddles can be imagined. From the late 1940s onwards flank girths also became more commonly used. In the early years cruppers were essential on bucking horses, for the location of the girth strap mount along the tree, which largely determines the stability of the saddle, is always a compromise. If placed too far forward it exerts insufficient influence over the rear of the saddle. Consequently, when a horse arches its back while bucking, the rear of the saddle lifts free and only the front has contact. In this situation, and especially with immature, round-backed horses, which have little wither, the saddle can slip forward very easily. Conversely, by placing the girth mounts further back along the tree, the girth is located further back along the rib-cage and thus, although the back of the saddle is now held down more firmly, the saddle is more likely to roll. Accordingly girths were set reasonably well forward and the saddle held in position with a crupper. Cruppers, however, can be dangerous to apply. They may also rub raw the skin under a horse's tail, thus making it irritable and more likely to buck. By mounting a second girth six to seven inches further to the rear, the disadvantages of a crupper are removed and the advantages of both girth positions realised. The flank girth, also known as a balance girth, is mounted on two straps that come off the tree to a ring. The two girths are connected by a small strap running along the belly, thus preventing the flank girth slipping backwards into the horse's flanks and causing it to buck.

In the early 1960s Hill, realising the shortcomings of most earlier saddles, designed his now famous Barcoo Poley for big northern stations:

in order to overcome the problem of conventional stock saddles hurting the horses' backs. Normally, during the wet season, saddles were counter-lined and got ready for the next season. Newly counter-lined saddles could be put on horses fat from their spell without any problems. Within three or four weeks the horses would lose so much condition that the saddles would sit lower and gall them on their withers ... [The old style saddles] were narrower in the pommel and the head of the Barcoo was built one and a half inches higher than any saddle before, so it would clear high withered horses, or horses in poor condition.[335]

Hill also replaced the usual doe and horsehair stuffing and check lining with heavy felt panels[336] to overcome the problem of poorly stuffed and ill-fitting saddles. Accordingly, new owners frequently 'paid for the Barcoo within a few years, compared to having to have the saddle counterlined each year.'[337] Furthermore, in a radical departure from tradition, he used solid, one-piece moulded fibreglass trees. Hill is dismissive of wood and steel trees, claiming they were:

never a good tree. Half the trees were cracked and broken before the saddle went out and onto a horse from all the tacking. Half of them would be split ... [and] would only fit the horse about three inches back from the gullet. The rest was anything but fitting ...[338]

Hill also argues that handmade traditional trees were all different, even if made to a pattern. Today, however, a rider can purchase several saddles of the same model based on a moulded tree and know they will be identical.

The question of the development of the Australian stock saddle is actually far less straightforward than would appear from this brief account. Claim, counterclaim and ill-informed opinion have frequently muddied the argument, and over the years the debate has often become acrimonious. Perhaps the best example of this is the vexed question of John Julius (Jack) Wieneke's contribution. R M Williams, for instance, states that:

> around the turn of the century, and about the time I was born, there was a man making saddles for the ever-critical riders of the brigalow. He [Jack Wieneke] lived and worked in Roma in the west of Queensland. His saddles were different. He altered the ideas that came from England with the immigrants, and developed a saddle more suited to the galloping scrub runners of the brigalow.[339]

Much has been written about this man. The Stockman's Hall of Fame features Wieneke exhibits and claims not only that 'almost every modern Australian stock saddle is derived from the original style of the Wieneke,' but also that he was the 'Father of the Australian Stock Saddle.' Similarly, in an article by former stockman Jack Gardiner, he is described as a saddler from Roma who:

> first achieved prominence through his athletic feats where he shone as an outstanding runner and jumper, but it was as a saddler that he was to become most famous.
>
> At 13 years of age he was apprenticed to the local Roma saddler Charles Arnold, from whom he bought the saddlery business in 1885 when he was aged 22. He soon established a reputation for the outstanding quality of his work. The fame of Jack Wieneke grew after he took one of his saddles to the Queensland town of Mitchell and put it on an outlaw horse that had never been ridden and then rode the horse with ease. The story spread quickly and he and the Mitchell Break as the saddle came to be known, had earned an enviable name among the horse riding community.

According to Gardiner, Wieneke then shifted to Brisbane where Lance Butler, the manager of the newly established British firm Edward Butlers, seized the opportunity to use his name on their products.

> All the saddles were manufactured under the supervision of Jack Wieneke and were top articles, but a disagreement brought about the breaking of the partnership. Butlers legally retained the right to use the Wieneke name so all the saddles Jack made from that time bore the stamp 'A Genuine True to Label Jack Wieneke Himself' and his signature.

Gardiner goes on to claim that:

> two of the prominent makers of top grade saddles Syd Hill and Schneider were both apprenticed to Jack Wieneke and the saddles they made were heavily influenced by their time with the master craftsman.[340]

Dealing firstly with the assertion that Wieneke is the father of the stock saddle, mention must be made of W H L Ranken's description of a stockman's saddle almost identical to the present-day one.

> His saddle is peculiar. It is very strong and heavy, say 24 pounds. It has enormous knee pads, very high up and far forward; the cantle far back and high, the seat very broad, and the pommel very low ... It is admirably adapted for rough country ...[341]

Clearly, the stock saddle as we now know it was developed at least twenty years prior to Wieneke's product. Indeed, it was in use when Wieneke was only about ten years old. Thus, despite the claims of Williams and others, the Australian stock saddle unarguably did not originate with Wieneke.

Renowned eighty-year-old fourth generation saddler Jim Hill, of Syd Hill Saddlery, who is almost certainly the last surviving prominent saddler to have been working in the industry in Brisbane during this period, dismisses the remainder of these claims. In the first place, he states that when Wieneke came to Brisbane he was employed by Lance Butler on repair work only. As to the statement that Wieneke supervised the construction of Wieneke brand-name saddles at Edward Butlers, Hill states emphatically that:

> all Wieneke saddles were made by my father, Syd Hill, and branded Wieneke. Wieneke never made any of them. He came to Brisbane [and] ... Edward Butlers, an English firm in Queens Street, were looking for someone to advertise and gave Wieneke a job. He was very well known inland as an athlete. Lance Butler wanted to advertise and sell the Syd Hill saddle, but my father wouldn't agree. He wanted to handle them on his own. So Butlers advertised the Jack Wieneke saddle throughout Australia at a cost of 3000 pounds, which was a lot of money in those days. That's how he got his name. The only saddle he could make when he was at Roma was the little old saddle with the flat back and the little knee pads right on your knee. He never made a big knee pad saddle in his life. They were all the Syd Hill type of saddle.[342]

In response to the claim that when Wieneke separated from Edward Butlers he began making his own saddles, Hill states that these saddles were made by Harold Worthington and later by Joy Scott, and sold by Jack Sims. Furthermore, Hill claims that the Australian firm Butler Brothers, which took over the Wieneke trade name from Edward Butlers in about the mid 1930s, produced the Mitchell Break along with other saddles under the Wieneke trade name. As Wieneke never worked for Butler Brothers, being approximately seventy years of age and semi-retired by then, Hill is emphatic that Wieneke never physically made a modern saddle while he lived in Brisbane. In short, he contends that all the saddles produced during that period which bore the differing variations of Wieneke's name were the work of Syd Hill Saddlery, Butler Brothers and Jack Sims. And finally, Hill denies categorically that his father was ever apprenticed to or influenced by Wieneke. After all, Hill senior's father and grandfather had been saddlers in Brisbane since the 1850s, and were the proprietors of the largest firm in Queensland.

Yet despite Hill's first-hand observations, myth has now been enmeshed with fact for over seventy years and the matter almost certainly will never be resolved to everyone's satisfaction. What is beyond dispute is that such developments aided riders in Australia and the US. Mora certainly has no reservations, stressing that:

> the old-timers, with their slick forks and long stirrups, had an unquestionably harder problem to keep glued to the hurricane deck on a volcanic bronc ... It doesn't take one of particularly keen mentality — just half normal will do — to realise that there's no comparison possible as far as the saddle advantages go. The moderns have the edge. Well, you'd scarcely call it an edge, it's plain 100% advantage.[343]

While it is clear that riding saddles did not change fundamentally, pack-saddles underwent radical changes. Being used exclusively to convey equipment, they were inherently different in design from riding saddles. Although information on early British pack-saddles is not available, it is almost certain that in Australia cavalry riding saddles, which have a large tree with several buckles and attachment points on either side, were mainly used in lieu of pack-saddles until at least the 1840s. During Thomas Bannister's 1830 exploration from Perth to Albany, for instance, his supplies were tied onto the riding saddles used. They not only frequently came off, but also made the saddles move. Accordingly, the horses soon galled along their backs and six days were lost through having to rest them. As the expedition was critically short of provisions, they were fortunate to reach the south coast. In a strongly worded report, Bannister asserted that future government expeditions should be adequately equipped.

> I would beg to recommend the necessity of having some description of pack-saddle for such expeditions, with a breast strap to prevent the load shifting back in ascending steep places and a crupper to prevent its pressing forward in descending. Our horses' backs were more injured in consequence of the constant shifting of the load than from the positive weight, [and] at this season of the year you are obliged to carry waters which of course adds much to your loads, [so] the only remedy is to be prepared with proper pack saddles.[344]

The illustration of Edward John Eyre's arrival in Albany in July 1841, after his gruelling expedition along the southern coastline, also portrays military saddles.

Even when British pack-saddles were available, scarcity meant they were often a poor fit. Western Australian Surveyor-General John Septimus Roe's safety and efficiency were compromised during his 1848 exploration of the southern regions of Western Australia, as most of his pack-saddles were originally constructed for ponies and never fitted correctly.[345] Consequently, although hard-pressed for supplies, he frequently had to rest his badly galled and exhausted horses.

The first pack-saddles designed specifically for Australian conditions were made by Augustus Gregory for his 1855–56 North Australian Expedition. Drawing on his extensive bush experience as a surveyor in the Western Australian Lands and Surveys Department, Gregory constructed a light, strong saddle

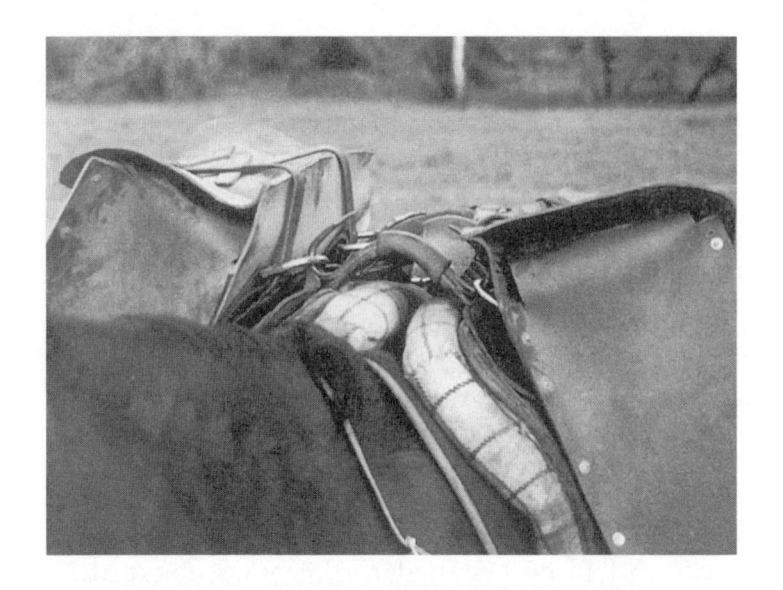

Plate 45: A Gregory pack-saddle showing attachment hooks.

which allowed air to circulate along the horse's back, so helping to prevent excessive sweating and galling. The specifications for his pack-saddle have scarcely changed in the ensuing 140 years.

> The pack saddle made under the direction of Mr Gregory consisted simply of two boards of Australian cedar, about 20 inches long to seven broad, inclined at such an angle as to sit fairly on the horse's ribs, and at such a distance from each other that the spine would remain uninjured between them. These were connected by two stout bows of iron, one and a half inches broad by one inch thick, arching well clear of the horse's back, and having on each side hooks firmly riveted into them for the suspension of the bags in which our provisions and etc, were stowed. [See Plate 45.] The crupper was buckled around the

aftermost bow, and the straps for the attachment of the breasting, breeching, and girths were screwed on the outside of the cedar planks ... A pair of pads, sufficiently large to prevent not only the saddle but also the packs chafing the horses, were attached to the boards by thongs passing through the holes bored in either end, so that upon occasion we could easily remove them to re-arrange the stuffing, and tie them again in their places. A thick felted saddle cloth was invaluable as an added protection ... The bags were of stout canvas, as wide as one breadth of the material ... no other fastening was used, so that if a horse fell in the rugged mountain paths, or in fording a rough swollen torrent, it was an advantage to him to shake off his bags at once ... Nothing whatsoever was allowed to be fastened to the bows above the suspension hooks; indeed there was a general order that the horse should carry nothing that was not contained in the side bags ...[346]

The strength and durability of Gregory's pack-saddles was clearly demonstrated when his horses bolted after the schooner *Tom Thumb* fired a farewell salute at the Victoria River depot in January 1856. After the horses had been tracked down and driven back to the base camp, Gregory noted with some pride that:

> a few of the straps of the colonial-made pack-saddles had given way, but there was no other damage done to them; but the English-made saddle was shaken to pieces.[347]

Not surprisingly, Gregory had a further thirty-one pack-saddles made to the same design for his Leichhardt expedition.[348] Thereafter riding saddles were gradually replaced by Australian pack-saddles. William Landsborough's journal, which was written in the Gulf country in the early 1860s during his search for Burke and Wills, mentions that his party stopped very early one morning, shortly after setting off from the Gregory River:

> to adjust a pack on a riding saddle. The other pack-saddles were constructed on Gregory's principle and required less adjusting.[349]

Modern pack-saddles differ from Gregory's only in that the padding is integrally related to the wooden part of the tree, having the leather facing wrapped around the wooden planks. Some further refinements have also been made through the development of patterns specifically for mules. Another development has been the use of rectangular leather saddle-bags in place of the less durable canvas bags. Their low attachment points reduce both their position and the horse's centre of gravity to at least that specified by Gregory. A large leather surcingle is also now used to completely encircle the horse and saddle-bags (Plate 46). When leather bags and surcingles were introduced is not known precisely, but a drawing from Alexander Forrest's expedition in the King Leopold Ranges in the Kimberleys in 1878 shows they were in use at that time (Plate 47).[350]

Curiously, despite the isolation of most of the pastoral region and Australia's small population, there have never been marked regional differences in Australian saddlery. By contrast, Mora outlines not only an east–west difference in United States saddlery, but also a number of subtle variations arising from differing usages, preferences and horse types. Certainly US saddlers and horsemen were considerably more innovative and more willing to experiment than their Australian counterparts. For example, cowboys who held lassoed cattle on the saddle horn used only a single girth, while ropers who attached the end of their lariat to a heavy breastplate used two girths to carry

Plate 46: A packhorse complete with a Gregory pack-saddle, saddle-bags, britching, breastplate, surcingle and Condamine bell.

the strain more safely. Furthermore, riders whose mounts were pot-bellied and a bit weedy also used a double girth. In this case the front girth was kept slightly looser than normal, and a thinner girth (strap) used to prevent galling. Another difference was that, whereas Australian stock-saddles had fixed girth mounting points, irrespective of whether single or double girths were mounted, US saddlers routinely varied the placement, constructing standard rigs, and variations forward of this known as five-eighth, three-quarter and seven-eighth rigs.[351]

Australian stirrups and stirrup leathers have also shown little variation. The stirrups were traditionally of the two- or four-bar British pattern, with wooden and metal oxbows a later introduction. Similarly, stirrup leathers were almost always a single strap of good quality leather one and a quarter to one and a half inches in width. Mora, however, outlines a variety of stirrup types, ranging from wooden and steel oxbows and circular hoops of iron, to simple circular pieces of wood about an inch thick, which were used by Mexicans. The Mexicans' stirrups had only a small hole for the stirrup leathers and a further small hole for the rider's toes, for Mexicans did not ride with their feet hard into the stirrups.[352]

The materials used in the fabrication of girths show a similar pattern. A wide range of materials was used in the US, including 'cotton, fishcord, mohair, horsehair (preferably mane), canvas or leather.'[353] By contrast, leather was used almost exclusively in Australia until the 1940s, when cotton cord girths became popular after they were found to reduce galling. This

Plate 47: Alexander Forrest's packhorses on the 1878 De Grey to Port Darwin expedition.

initial reliance on leather was due to the almost complete lack of roads and railways in northern Australia and consequently extremely high cartage costs. Leather, which was plentiful on cattle stations, was the logical if not the only choice.

The securing of girths has shown similar uniformity, with only two major methods being employed in Australia. Southern saddles have traditionally used tree-mounted girth straps that fit through buckles on the girth. Pastoral saddles, however, have for the most part used a supple leather strap attached to the saddle. This strap is laced through a ring on the end of the girth and then tightened and tied off. The main advantage of the latter system is that girth straps are not worn out by the immense amount of work pastoral saddles are subjected to. In addition, more leverage can be exerted through the pulley-like system, thus allowing the saddle to be tightened easily.

While nearly all Australian breakers and stockmen traditionally used a heavy snaffle bit with cheek bars, US horsemen were divided on an east–west basis. The eastern group generally used curb bits while the western group often mouthed their horses in a hackamore before changing to a Spanish spade bit.[354]

Finally, a marked national difference can be seen in the adornment of saddlery, notwithstanding that both Australian and United States horsemen were descended in large part from British stock. Almost without exception Australian equipment was strictly utilitarian, having no adornment whatsoever, whereas US saddlery was frequently embellished with inlays of brass, copper or silver, and leather engraving and embossing. Although Hollywood's depictions of a working cowboy's outfit need not be taken too seriously, it is undeniable that US cowboys were more flamboyant.[355] Their stirrups, for instance, were often covered with leather facings called tapaderos, or taps. Frequently these were lined with sheep pelt and were extremely useful in keeping the rider's feet dry and warm in the severe winters experienced in many parts of that country. Tapaderos were not only functional, however, for they were often embellished with inlays, engravings and metal buttons. Furthermore, they were frequently cut so as to hang downwards in a graceful sweep, with examples as long as twenty-six inches being recorded. Mora believes that, when well made, they could be 'the smartest, handsomest things you'd ever want to see,'[356] and concludes by outlining what he considers aesthetically pleasing.

> My notion of what's tops is an old-time California bit shank on rustable steel, with a good silver concha and inlays; good-sized conchas on lightweight round leather headstall; all-silver string conchas on the saddle and taps, and initials or brand in silver at the back of the cantle. That's all for silver. However, I do admire fine flower embossing on the leather, preferably that small-scale wild-rose pattern. When saddle leather is just encrusted with this type of design well done, and has taken on the colour and shine of constant use, and is set off with dashes of silver as already indicated, I don't think there is anything handsomer in the cowboy wardrobe.[357]

Certainly most Australians expressing such sentiments would be accused of being tongue-in-cheek.

Horsebreaking

It is somewhat ironic, given our bushman legend, that many Australians have no conception of the process of horsebreaking and appear to have a hazy idea that, perhaps not unlike vehicles, horses are inherently rideable. But horsebreaking is a highly skilled task which involves mouthing, riding and educating. The first of these involves the animal being taught to respond to directions to slow down, stop, turn left or right and back-up. These instructions are transmitted via reins to the bit in its mouth. Riding simply entails working the horse until it has stopped bucking and become suitable for a less experienced rider, while educating encompasses teaching a number of skills to make it a more useful and tractable working horse.

The British technique of horsebreaking, which would have been the model for our earliest stockmen, has a long tradition and is based on certain premises. The most important of these is that British horses, as a result of being hand-fed, often shedded through winter and usually confined in small paddocks or enclosures, are in most instances not only accustomed to the sight of humans but also well-handled from an early age. Consequently they are tractable and even-tempered. Another factor is that these horses have a market value that reflects the cost of the feed and wages expended on them. In short, they are a valuable commodity and treated as such in the breaking-in process. British horsebreakers, for example, rather than merely fitting a standard bit at the beginning of the breaking-in, frequently use special mouthing bits. These have dangling keys of metal that the horse can play with, with its tongue, to avoid boredom. Other subtleties include bits which incorporate inserts of copper. These are intended to stimulate the flow of saliva and thus prevent the problems that can arise from dry mouths.

In this system of breaking, the horse is often left to roam around the yard wearing only a bridle, merely to let it 'get the feel of the bit.' The procedure may be carried out for some hours each day for up to a week. Then a roller, a specialised leather strap which fits around the horse's girth, is fitted loosely and gradually tightened. When the horse is comfortable and accepting the roller, side-reins are added. These are adjustable straps connecting each side of the bit to the roller which allow the horse to be gradually reined in shorter, thus restricting the extent to which it can lift its head. By this technique, known as tying-down, the horse learns to accept pressure on its mouth without throwing its head up, which is essential when riders begin to ask the horse to slow or stop.

Once the roller and side-reins have been successfully accepted, long reins are fitted. These twenty-five to thirty foot-long webbing or leather straps pass through a buckle on each side of the roller and are clipped to the ring of the bit on the same side. By this arrangement the horsebreaker can stand in safety, well clear of the horse, and gradually teach and refine stopping, turning and backing. Eventually, when the horse is suitably controllable from the ground, it is ridden. At this stage it is saddled and the rider, after mounting quietly, gently urges the animal forward. If the preceding mouthing has been satisfactorily carried out, the horse usually steps forward in a steady, controlled fashion and soon begins learning more complex skills. By this carefully established sequential process, horses in Britain are taught quietly, and usually with little coercion, to accept riders and to cooperate and perform in a highly sophisticated manner.

Overall, there is an expectation, indeed almost a requirement, that they will be broken in superbly. As a result, practitioners take considerable pride in their work. However, because of the unhurried nature of the process, during which no stage is commenced until the preceding one has been thoroughly absorbed, it is a time-consuming and expensive method. Indeed, breaking in horses by this traditional approach can take ten to twelve weeks.

Although the equipment used in the US in the 1800s often contrasted strongly with that in Britain, the emphasis was similar.

> In the old days ... those amansadores [breakers] had lots of savvy, plenty of patience, heaps of mañana, and a small value on time ... A colt would be worked six months or a year with a hackamore; then he'd be made to pack a bit for a while extra without any reins attached to it; and then he'd graduate into the two-reined class ... That is where he'd get his thorough and final schooling.[358]

By the early 1900s, however, economic constraints led to a reduction in the time and skill expended on breaking and educating horses. According to Mora, 'the modern, streamlined efficiency business [meant] ... things must be done in two months that took the old-timers a year or more.'[359]

Such painstaking care and pride were not evident in the pastoral regions of Australia, where the British premises and practices were simply not applicable. Firstly, although large numbers of horses were bred on these vast properties, they often had little commercial value because of the costs involved in mustering and transporting. Accordingly, stations could not afford to spend too much on breaking-in and generally paid professional breakers on a contract rate, whereby they had to complete a set number of horses each week to make the job worthwhile. Secondly, because of their low value, many station horses were untouched by humans as young animals. The usual station practice of not breaking in horses until three or four years of age or even older, due to immaturity, compounded this early neglect.[360]

These factors, combined with poor mustering practices, meant it was not uncommon to see unhandled stallions of up to eight years of age run into the yards for breaking. Not only were such horses almost valueless, they were also quite wild. Indeed, whereas British and southern Australian horsebreakers generally could walk up to and catch horses with no particular difficulty, most unhandled station horses responded to a direct approach by flight. If cornered in a yard, their normal response was to turn away and kick ferociously with both hind feet. By this means they could fracture a breaker's arms and legs, crush ribs, smash facial bones and even kill him. Archie Skuthorpe, cousin of the renowned buckjump riders Lance and Dick Skuthorpe, died at Auvergne station in June 1918 when he was kicked in the head by a horse. At the time Skuthorpe was:

> breaking in colts along with the station blackboys. One of the boys had saddled a colt preparatory to riding him for the first time, and, as the crupper was loose, Mr Skuthorpe tightened it a bit. After fixing the crupper he turned and walked away when the horse gave a sudden bound forward and lashed out with both hind feet, kicking the deceased on the shoulder and side of the head. After the accident Mr Skuthorpe was able to walk, with assistance, to the house, about a quarter of a mile, but never regained complete consciousness, and died about ten o'clock the following morning.[361]

Alternatively, these horses could bite, or rear and strike savagely with their front feet. Author and former Pilbara drover Thomas Cockburn-Campbell provides a graphic description of the latter, when describing how he and his partner Billy hobbled a mob of brumbies they had trapped in a yard on the edge of the Great Sandy Desert during the 1930s. They firstly roped each horse from the top rail, then dragged it up to a rail fence. By reaching through the rails they were eventually able to hobble all the horses, but it was very dangerous work for 'they struck like fury and there were a lot of near misses.'[362]

As a result of these factors, and because in many instances the horses being broken were sent straight to mustering camps, put into full work and ridden by proficient horsemen, little time was wasted. Pride was taken in the speed and cheapness of the process, and near enough was definitely good enough. Former cattleman Pat Underwood clearly recalls how:

> the old style of breaking was to get him and choke him until you could get your hands on him ... [Some] people didn't believe in long reining. They'd jump straight on and ride them straight away ...[363]

Former Territory breaker and stockman Reg Baker had similar experiences.

> A lot of fellows used to just jam a saddle on and ride them in the yard once and hand them over to the people in the stock camp ... They were broken in dreadful circumstances.[364]

In short, as Campbell recounts:

> Breaking in horses was a streamlined job then — we spent only one day on each horse. The methods were rough and ready and many horses became outlaws through lack of tolerant training.[365]

This is not to say that all breakers were alike. For over 150 years a considerable number of stations relied on the services of skilled professional breakers who handed over well-mouthed and well-handled horses. Indeed, some horsebreakers in these regions, such as Jack the Quiet Stockman, described by Jeannie Gunn of Elsey station, were renowned for their patience, gentleness and skill.[366] Nevertheless, the style and tradition of breaking that developed in pastoral regions was the antithesis of that in Britain and the US. Not surprisingly, by the mid 1800s many British commentators were scathingly critical of Australian breakers' standards. In *Geoffry Hamlyn*, when asked if he could break horses the British youth Dick replied heatedly:

> I'll break horses against any man in this country — though that's not saying much, for I ain't seen not what I call a breaker since I've been here.[367]

Similarly, in 1874 Staff Veterinary Surgeon and remount agent W Thacker reported that one of the major faults of Australian horses exported to India was that they bucked badly. Thacker had no doubt that:

> 'buckjumping' is entirely the result of the hurried and imperfect mode of breaking so generally practiced with young horses intended for the Indian market. Sufficient time is not allowed for instruction, submission being enforced, and not taught, and when opportunity arises, opposition or resentment is shown.[368]

While this is a contentious topic, few would argue that pastoral breakers showed considerable ingenuity, perseverance and courage in developing the range of innovative techniques and equipment necessary to deal with station horses. The first step was to muster them (often from a hundred-square-mile paddock) into cattle yards. Once in the yard a proportion of these horses could be caught from the ground, although sometimes a long pole was used to push an animal's hindquarters into a corner, to make it face up to the breaker and reduce the risk of him being badly kicked. Other techniques devised to help catch frightened and dangerous horses included the use of a stockwhip. Most stockmen have used a whip on a horse's hindquarters to make it face up, for 'they never forget ... if you give them a flick or two with the whip.'[369] H M Barker also mentions how, as horses circled renowned Gulf breaker Jack Brady in a yard during the early 1900s, he would casually throw hand-made halters over their heads. Brady rarely missed and would then pull them around by the halter shank.

More difficult horses were roped as required. They were then pulled up short and tied hard against a high rail fence. In earlier days roping was frequently carried out with the aid of a roping-pole. The rope, with a noose at one end, was draped along a long pole which allowed the horsebreaker to get close enough to flick the noose over the horse's neck. This practice gradually faded as breakers became proficient at roping. When roped, the terrified animals frequently struggled violently, pulling the noose tighter and tighter, until they became unconscious from lack of air. This process, known as choking down, not only taught horses to respect humans, but also never to pull back and snap ropes. For the uninitiated, however, seeing a horse choked down is an alarming sight. The poor animal roars and gasps as it strains desperately to get air into its lungs and its neck is compressed into what seems a quarter of its normal size. Staggering and wobbling increasingly, it eventually falls unconscious. At this stage the rope has to be loosened quickly in order to save the horse.[370]

Once a horse was snigged alongside a rail fence and reasonably settled, some breakers restrained it with hobbles, or hobbles and a side-line. This enabled the breaker to handle the horse with a degree of safety as he worked to overcome its fear of humans and nervousness at being touched. It also prevented it from charging off. In 1928 Charlie Schultz of Humbert River station was advised to use side-lines when his freshly broken horses were proving difficult to catch. Once they were fitted, the horses fell to their knees if they moved too quickly in their efforts to escape and Schultz found he:

> only needed to do that about three or four times before they'd realise that they couldn't get away. Then you could just walk up to them quietly and rub them down on the forehead, and put a bridle or halter on.[371]

Alternatively, breakers used a spider, which consisted of four pieces of chain radiating out from a central ring, with a hobble strap passed through the outside link of each piece of chain. By buckling the individual hobble straps around each pastern, in a potentially extremely dangerous operation, all four legs are linked together. The horse is then completely under the horsebreaker's control. Yet another method was to use a collar rope. A heavy rope was passed twice around the horse's neck and tied, then a second rope looped around a back pastern and passed through the collar rope. The back leg was then pulled clear of the ground and the leg-rope tied to the neck-rope.

Once the horses were suitably restrained, breakers began handling them. Some used a long thin stick, which allowed them to stand back and touch the horse. Others used a bag, which they rubbed, slid and flapped against the horse in a process known as bagging. In the latter case the bag was often tied to a stick, particularly when attempting to rub the back legs of nervous horses or known bad kickers.

At this stage horses were also taught to ground-tie, meaning that they would stand stationary if their rider left one rein trailing to the ground after dismounting. The reasons for ground-tying are obvious: the rider does not have to tie the horse on each occasion (for which trees are not always available), nor hold it by the reins all the time. Ground-tying is relatively easy to teach. The horse is tied to a sandbag of a weight such that he can move it only with considerable difficulty, and allowed to drag it around for some time. Eventually the horse comes to associate a rein or rope hanging from its head collar to the ground with limited freedom and can be left loose with confidence. Unfortunately, once a horse or mule discovers it can in fact move off, it can become extremely cunning and difficult to catch. Campbell cured a rogue mule, Ikey, of the habit in the late 1930s by hobbling him and tying a long rope to the hobble chain. The rope was then passed out between Ikey's hind legs and tied to a nearby tree. After being struck violently:

> Ikey shot off like a rocket. He used up the slack in the rope, and then tumbled over and over. Great gasps of air were expelled from his open mouth as he hit hard. We followed this procedure three times until Billy could walk up to Ikey, hit him over the head ... and he would not move.[372]

As soon as the handling was completed, breakers placed a bridle and roller on the horse, which by now was often running with sweat and 'scouring' (defecating) copiously from fear. To ensure the horse stood relatively still and allowed the breaker to fit the equipment with a degree of safety, it was often snigged up tightly to a centre-pole in the middle of the yard before work commenced. Alternatively, the horse was run into a drafting race and, by standing on or outside the rails, the breaker could apply the necessary equipment with a degree of safety. Yet another method was to use a quiet and well-educated older horse as a shield, by which means a breaker could place equipment on very fractious animals.

To protect his face from being seriously injured by a terrified horse which objected violently to having its ears bent as the bridle was slipped on, a breaker often used a halter-bridle. These consist of a plain halter, which can be fitted more easily and safely, and a bit attached by clips to the lower buckles on

*Plate 48: A horse being mouthed in the Kimberleys,
head tied down and strips of saddle-cloth trailing, 1960s.*

either side. Before the halter is fitted, the clip on the side where the breaker is standing (the nearside) is undone, so that the bit hangs down from the offside buckle. The halter is then fitted, the horse's mouth opened by a judiciously inserted finger or two, and the bit drawn across and up into the mouth and re-clipped to the halter buckle on the nearside.

Once the horse was fitted with a roller and bridle, it was left to walk around for some hours to become accustomed to the feel of the equipment, especially that around its girth. To help overcome touchiness, breakers usually left ropes or pieces of cloth trailing from the roller (Plate 48). While initially the terrified horse kicked ferociously every time it was touched, it gradually became desensitised.

During this time the horse was also tied down. Plate 49 shows a team of horses at Flora Valley station part-way through

Plate 49: Horsebreaking at the Flora Valley station yards, 7 December 1925.

the mouthing, with rollers and side-reins fitted. This step was an exacting one, for if the animal was tied down too severely at first, it could panic and rear over backwards. The side-reins were therefore tightened gradually and judiciously. In doing so, a horsebreaker placed himself in danger, as a horse restrained in this manner can rear, thrashing its head sideways and hitting the breaker an extremely heavy blow in the face. Subsequently, the slightest movement or adjustment of equipment is often sufficient to frighten a nervous horse, which can strike, kick backwards, or cowkick sideways in fear.

Cowkickers have a much wider reach, as they can strike in an arc anywhere from the immediate region of the hind leg to forward of their front leg. Campbell recalls that, when mounting one particular horse:

> you had to rein him up tight on the nearside, stand in front of the shoulder and swing fast into the saddle. Even then he would let go three fast and furious cow kicks before you hit the saddle. Getting off, you had to

rein in hard, and swing towards his head quickly. He would cow kick then, too, in a very meaningful way.[373]

Although not as dangerous as a full-blooded backward kick, cowkicking can still be injurious. On one occasion Simpson was breaking in a packhorse and had it hobbled and side-lined. While restraining the animal and covering its eye, he asked his offsider Ron Condon to gently remove the rear hobble strap. The horse suddenly lashed out at Condon, who was bending towards it, striking him on the forehead and knocking him unconscious.[374]

Once the horse had come to accept being tied down, long driving reins were fitted and a minimal amount of time spent in teaching the rudiments of stopping and turning. While horsemen such as Baker and Underwood took especial care with long-reining, to ensure their horses ended up with light, responsive mouths, mouthing was not generally a strong feature of station breaking.[375] Time constraints and perhaps an element of bravado frequently led skilful and not-so-skilful station breakers to complete from the horse's back much of the mouthing traditionally done on the ground. Thus there were very wide variations in the standard of work. In one instance Jack Brady, whom Barker considered 'the best of the lot,'[376] was mouthing and then riding horses twice in the yard. Barker and another youth then rode them out from the yards into the bush for two hours each day. His description of Brady's handiwork is particularly revealing, for he not only states that this was exhausting work, but also mentions how on leaving the yards each horse:

> would think it had escaped [and] start off at a gallop ... and then would begin the long tussle of pulling it up to a walk ... The hardest work was cantering about lugging the horse this way and that, always against its will, until it learnt to come around ...[377]

Clearly, by traditional standards, these horses had no mouth whatsoever. They had no idea of turning or stopping and were being taught when too tired, after hours of riding, to fight any more. Schultz appears to have adopted much the same approach as Brady. In his description of the initial breaking-in on Humbert River station in 1928, he makes no mention of tying the horses down or driving them. Furthermore, even though he was 'very pleased' with the job, he admits that the horses were hobbled the first time they were ridden out and that an Aboriginal rider was always present, otherwise the horse would 'go bush! He'd just gallop away you see, and not being used to a bridle, you couldn't pull him up.'[378] Campbell's hair-raising exploits, although humorous, exemplify the shortcomings of the mouthing process. On one occasion his horse bolted 'hell for leather' along a river bank, leaping tributary creeks which bisected its path and passing under perilously low hanging branches, then swerved across the riverbed and clambered up the other bank before smashing through thick mulga scrub until, heaving furiously, it could go no further.[379] Unfortunately, while this approach was functionally adequate for at least 150 years, many horses and men suffered badly or were killed in the process.

Inadequate initial mouthing was not the whole problem. The situation described so well by Barker is instructive:

> The horse that leaves the breaker badly mouthed rarely ever stops throwing its head about, sawing at the bit, and holding its nose high and forward when asked to make a turn ...[380]

These animals were almost certainly suffering from teeth problems. Horses masticate their food obliquely, and gradually wear extremely sharp edges on the outside of their upper teeth. Consequently, when a heavy-handed rider attempts to drag a horse around or pull it up, the bit forces the insides of the cheeks onto these very sharp points, which frequently causes severe lacerations. Naturally the horse attempts to move away from the source of pain by throwing its head in the air or going backwards. Alternatively, it may lock its jaws onto the bit and run round in circles. Even if its teeth are not overly sharp, the horse may be losing caps (its baby teeth) as a three-year-old, have particularly sensitive grass-seed injuries inside its mouth, or be suffering from other problems.

Whatever the cause, horses should ideally not be ridden without having their teeth rasped smooth, and this is a common aspect of horsemanship with well-handled and valuable animals. Horse dentistry has never been a feature of station breaking, however, either from lack of knowledge or because of the practical difficulties involved. Rasping a horse's teeth entails standing in front of the animal and, given station horses' propensity to strike, the operation is simply too dangerous unless a crush is available.[381] Some station breakers adapted their techniques in an attempt to overcome the problems of lacerated mouths and unresponsive horses. Unfortunately, once horses so affected become sulky and intransigent, it can be very difficult to mouth them satisfactorily. Indeed, many horses have been very badly mouthed or even ruined through not having their teeth correctly rasped, which in part explains Campbell's comment regarding turning horses into outlaws through lack of tolerant training.

No matter how well or how poorly horses were mouthed, once the breaker was satisfied, the next step was riding them. Unlike British and many southern Australian horsebreakers, pastoral breakers often faced a further struggle to saddle these horses. Frequently they became extremely nervous as the saddle-blanket and then the large, heavy saddle were gradually brought closer and lowered onto their backs. In many instances they bolted forward, cowkicked or reared to dislodge the saddle. As a result breakers learnt to stand well forward, even in front of the front legs, as shown in Plate 50, while manipulating the saddle and blanket into position. Often some form of restraint had to be used.

Plate 50: Ralph Speck, Bert Kite and an Aboriginal youth saddling a brumby at Cutta Para Corner yard, Cooper's Creek, c. 1941.

Apart from hobbles and side-lines, a variety of equipment was employed. The simplest was a twitch. This is an eighteen-inch length of axe-handle or comparable piece of wood, with a six-inch loop of light rope attached to one end. Breakers put their fingers through the loop and draw a handful of the horse's nose and lips back through, then turn the handle, thus tightening the loop down hard onto this sensitive tissue. The pain engendered is usually sufficient to make a horse stand quietly.[382] Care has to be taken, however, to stand near the horse's shoulder in case it strikes. If its front feet could be lifted safely, another option was to lift and fold a lower front leg hard back against its upper leg, until they lay adjacent. A stirrup leather was then wrapped around the folded leg and buckled firmly, so the horse was left standing on three legs. Should it then rear, buck or plunge in an attempt to avoid being saddled, it was usually chased along with a stockwhip, for it would tire quickly and then stand quietly. Another approach was to use a breaking tackle, which works on the same principle as a collar rope and leaves the horse standing on three legs.

Having saddled the horse, breakers were then faced with the question of where to ride it. In most instances they were working within the station's cattle yards. With fences normally about six feet high, such as those shown in Plate 51, these yards provided the breaker with some control by constraining a panicking horse and preventing it bolting for miles through the scrub. Once saddled, a horse could be allowed to buck itself out unfettered, or run riderless around on the end of a long rope attached to the head collar. This practice is known as lunging. Should the horse wish to buck it was encouraged, for it was expending energy that would otherwise be reserved for the rider. At least some of the roughriders employed in the Army Remount Section during World War I employed the technique. This group of several hundred horsemen, whom Banjo Paterson considered 'possibly the best lot of men that ever were got together to deal with rough horses,'[383] had the unenviable task of re-educating the scourings of the equine world so that fully-equipped troopers could mount them under fire. On one occasion the worst of a bad batch of horses, which General

Plate 51: Horsebreaking at Victoria River Downs, September 1941.

Royston insisted was to become his personal charger, was kept until last. As a precaution it was lunged for a while but, obviously knowing what was next, 'refused to take anything out of himself.' Accordingly, once mounted, it bucked violently with its 'head right in under his girth,' leaping and roaring all the while, and eventually ended in an irrigation canal.[384]

Strangely enough, Royston was adamant that he could handle the horse and told Paterson he would be very hurt if it was not reserved for him. Paterson felt Royston would be very hurt in other ways if the horse was reserved for him, but said nothing, for 'one doesn't say these things to a general.' Why Royston did not take up his claim to the horse is not known, but it was eventually used to provide buckjumping demonstrations for visiting English aristocracy. These 'were very popular ... and created a good impression that we were doing our job.' When asked by one dowager if he rode many of the outlaws, Paterson modestly replied, 'Only those that the men can't ride.'[385]

Despite these advantages, breaking horses within yards had its drawbacks. Occasionally, terrified horses smashed straight into the rails in an attempt to go through or over them, and breakers suffered serious injuries to their legs. Other hazards existed, such as cap rails, which prevent the uprights being forced apart by the pressure of crowding cattle. Cole once struck his head on such a rail when his horse bucked high as he rode out from the yard.

While out in mustering camps, where yards were rarely available, stockmen confronted with a horse they believed would buck soon became aware of the value of a patch of heavy sand. Loose sand not only bogs a horse down and reduces its capacity to buck as hard or for as long, but also provides a softer landing should the rider be thrown. Furphy's fictional character Tom Collins commented on this in *Such is Life*. Collins had just acquired a horse which he knew would buck badly each morning on first being mounted, and outlined why it was absolutely imperative he should travel thirty miles before establishing his next camp.

> I was aware of a sand-hill composed of material unstable as water; an unfavourable place for a bucking horse, and a favourable place for a man to dismount head foremost.[386]

As mustering camps were frequently adjacent to creek- or river-beds, riders were often well served in this respect.

Mounting nervous or poorly handled horses was a necessary skill which horsebreakers had to develop very quickly. Surprisingly, in almost all instances feeling a rider mount does not frighten a horse, whereas seeing him mount does. When this was realised, breakers developed a variety of techniques to block the animal's vision. Perhaps the simplest of these was having an offsider cup his hand over the horse's nearside (left) eye to obscure its rearward vision. Blindfolds were also used for particularly nervous horses. Using this technique, a horse's vision was totally occluded and it tended to stand quietly. So long as the breaker let the animal know where he was, by talking and physical contact, he was reasonably safe. Traditionally, blindfolds were hats, saddle-blankets, towels, neckerchiefs and jute bags. Some United States horsebreakers also resorted to a broad strap of leather which was attached to the cheekstraps on either side of the bridle and laid across the horse's face. This strap was pulled into position across the animal's eyes before mounting and, in a seemingly dangerous manoeuvre, the mounted rider leant forward and pulled it upwards and free of the animal's eyes.[387] In doing this the rider would be well forward and off balance, and at serious risk if the horse took fright and sprang backwards.

Another option was having an assistant grasp the horse's nearside ear and pull down hard. Not only did his arm obscure the horse's rearward vision, but the resultant the pain made it less likely to move. In Plate 52, in a somewhat unorthodox approach, not only is the rider screwing down the horse's ear, but he also has it hobbled and held with a rope. Sometimes bad horses required two assistants, each holding an ear and a bridle

*Plate 52: Bert Kite, clearly dubious about mounting,
Cutta Para Corner yard, c. 1941.*

cheekstrap. In his outline of Jack Brady in the early 1900s, Schultz provides a graphic description of the practice, known as lugging, as well as the extremely hard nature of the job and of the men involved. At this time Brady was suffering the legacy of an earlier very bad break to his leg.

> He'd yell — a savage old bugger he was — 'Hold that bloody horse's head here, hold him tight! Grab him by the ear or something.' And they'd say, 'I don't think you should get on that horse Jack.' He'd say to them, 'I'm all right. I'm as good as ever I was,' and he'd hit the bloody saddle and the bloody horse would fly away bucking.[388]

Holding a horse by the ear appears to have been more widely practised in the US. Mora often held both ears and swung his legs off the ground to increase the pressure on the horse — and the pain. Although assistants were frequently dragged across the ground, 'in general, we could wrestle them and get the job done handily.' Nevertheless, he is adamant that, although they 'rassled broncs of the mustang size ... I wouldn't try to ear down a 16 hand, 1200 pound part thoroughbred bronc,'[389] such as were prevalent in Australia.

Whether blindfolded or not, a nervous horse was always reined up very short to the nearside before a breaker mounted it, so its forehead was nearly touching his shoulder. By this means, even if the horse began to buck before the rider had fully gained his seat and offside stirrup, it tended to buck in a tight circle to the left. Not only did this help propel the rider into the saddle, but this technique also prevented the horse from swinging away and kicking backwards. Furthermore, bucking in a tight circle is not as effective for the horse as being able to buck straight ahead, unhindered.

Curiously enough, the riding of wildly bucking horses, which usually only occupies a minute proportion of the breaking-in, has long been seen as the heart of the process. In fact bucking is extremely physically demanding and very few horses can sustain it for even a minute. Those who are ridden out usually pull up heaving so badly from the exertion that the rider can feel their heart pounding between his legs. Apart from bad buckers and older, stronger horses, which recuperate quickly and are more than willing to recommence the next morning, most horses stop bucking after being ridden a few times. In fact the majority are capable only of a poor quality or even half-hearted attempt, known in Australia as pig-rooting in which they buck neither hard nor high. Others lack even that capacity, and merely leap and rear in a haphazard fashion. Nevertheless, some station horses could and did buck viciously on being ridden.

Furphy provides a wonderful description of bucking on the occasion when his hero, Collins, is artfully tricked into riding a particularly nervous colt. His fellow horsemen suddenly remember they have promised their wives not to ride colts for the current year, or profess a lack of ability. The final apologist has a boil on his buttocks 'which keeps him standing in the stirrups even on his own old crock ... [and] compels [him] to forego the one transcendent joy of his life.'[390] Consequently Collins finds himself astride the horse, nervously contemplating how its back is arched and its tail 'Ah! the saints preserve us! ... [is] jammed hard down.' His tormentors then insensitively ask:

>'Ready, Tom?'
>'Yes.'
>'You're sure you're ready?'
>'Yes.'
>'I think he'll buck middlin' hard ... '
>'You've got the off stirrup all right, Tom?'
>'Yes.'
>'I'm goin' to let the beggar rip.'
>'Go ahead ... '

But your voice is not what it ought to be, and the soles of your boots are rattling on the flat part of the stirrup-irons.

The chap draws the handkerchief from the colt's eyes, and walks backward. The colt catches sight of your left foot, and skips three yards to the right. In doing so, he catches sight of the other foot, and skips to the left. Then everything disappears from in front of the saddle — the wicked ears, now laid level backward — the black tangled mane — the shining neck ... they have all disappeared and there is nothing in front of the saddle but a precipice. There is something underneath it, though.

How distinctly you note the grunting of the colt, the thumping of his feet on the ground, and the gratuitous counsel addressed to you in four calmly critical voices:

>'Lean back a bit more, Tom, and give with him.'

'Don't ride so loose if you can help it, Tom.'

'Hold yourself well down with the reins, and stick to him, Tom.'[391]

'Stick to him, Tom, whatever you do.'

Ay! stick to him! Stick to the lever of a steam hammer, when the ram kicks the safety trigger! ... However, you have stuck to him for a good solid sixty seconds;[392] now, one of your knees has slipped over the [knee]pad, and your stirrup is swinging loose. Good night, sweet prince.

And away circles the colt, slapping at the bit with his front feet ... And away in chase go two of the chaps on their bits of stuff. Meanwhile, you explain to the other two that the spill serves you right for riding so carelessly; and that, though your soul lusts to have it out with the colt, a stringent appointment in the township will force you to clear as soon as you can get your saddle. Such is life.[393]

Not surprisingly, Australian breakers and stockmen developed a unique and highly effective style of riding They had to, for if thrown they often faced a long walk as their horse galloped wildly back to the camp. As Simpson notes, 'There were no pick-up men in the bush.'[394] Firstly, they rode with their feet well forward to guard against the horse stopping or changing course suddenly and to protect themselves against bucking. A rider has to pivot on his seat as a horse bucks, moving his legs forward and his shoulders back in order to maintain his centre of gravity. And the harder a horse bucks the more a rider's shoulders go back and his feet come forward. Indeed, poet and horseman Adam Lindsay Gordon, who was renowned for riding bad horses, was reputedly criticised by British judges for having his shoulders touch the rump of a bucking horse. Secondly, and what was probably most galling to British purists, many Australian horsemen learnt to ride with their feet hard into the stirrup and their toes pointing down. Although inelegant to classicists, this technique gives an immense advantage, for the rider is less likely to lose his stirrups at a critical moment.

> Their style of riding, however, does not exactly answer the usual idea of excellence; they have generally a long and loose seat, with the foot [hard] home, and the toe pointed to the ground in a line with the knee, and they seldom have a good hand on their horse, but not withstanding this they are very expert at sticking on under difficulties, and have a most astonishing knack of getting along very fast in broken country, and especially downhill, in which, perhaps, they are unequalled.[395]

Curiously, although over the years many horsebreakers and stockmen must have experienced emotions ranging from dubiousness to outright fear when mounting horses they believed would buck, little mention is made of this. Indeed, apart from Furphy's amusing description above, which was written in the late 1890s, there is almost complete silence on the subject until the rash of publications of the 1980s and 1990s. Presumably the reluctance to admit to such emotions arose from the obligatory requirements of stoicism and manliness. This absence is intriguing, for there is clear evidence that fear was recognised and accepted by horsemen. Certainly newcomers to the industry were warned never to say they could 'ride'. Tom Cole, for instance, was told as a green youth to reply only that he could handle a quiet horse, on the basis that:

> Anyone who goes on to one of these places and says he can ride needs to be in rodeo class or he's a mug. Don't get caught with that one![396]

Perhaps more startlingly, Campbell recounts how in the late 1930s his supposedly tough mate Billy, who selected what turned out to be the worst ride from a batch of three, broke down and cried from pain and fear when his horse toppled on him while bucking furiously.[397]

Mora provides the best account of nervousness in describing his general reluctance and eventual demise as a horsebreaker. As he points out, he was never:

> one of those big, HE, 'gimme a bronc for breakfast,' boys ... The truth is that some of my sourest recollections are the result of that bronc for breakfast stuff. Those bitter cold mornings before sun-up, trying to bolt a bait [breakfast] through chattering teeth, and (if you had to ride one of your rough string) casting a look over into the semidarkness where that equine criminal they called a broke pony was standing there quiet enough, his head down, his eyes staring cold, every hair on his body sticking up like a door mat, and a hump in his back that tilted the rear skirts of the saddle so's you could hide a hat under them. Forking a hostile hurricane deck when cold and stiff is not what I'd call ideal conditions for making a good ride. Well, I'll confess that after my first couple of years, all my bronc riding was done because it was positively necessary and certainly not from choice.[398]

Mora's final buckjumping ride came about after he had been badly thrown by a horse he had owned for years, but taken too lightly. On the following day he decided to retrain the animal and use it as a buggy horse, but perversely decided on one more ride to demonstrate to himself that he was as good as ever. As Mora went to put his foot in the stirrup, he realised something had changed forever.

> Well, it's hard to explain what happened. Something suddenly sparked through me that I can't describe; but I knew I was licked, thoroughly licked. I should have quit right then and there and followed that impulse, or

hunch, or whatever you might call it. But I didn't stop a second in my routine motions, and my foot went into the stirrup, I eased into the saddle, caught my off stirrup properly, and then Champagne's head disappeared as he made a couple of fast bucks ahead with a quick twist to the left. Any rider should have sat that out, for he really had not as yet got warmed up to his performance, yet I just poured out of that saddle and landed in a trash pile of ashes and barrel staves. Any farmer should have stuck those first few jumps ... My bronc riding days were over.[399]

Furphy and Mora aside, once the breaker had ridden a horse two or three times in the yard, it was taken out and worked through the bush for some hours each day for two or three days. During this time, once the horse had become tired enough to settle down and cooperate, it was taught the skills required when it commenced working cattle. These included learning how to sit down slightly on its hindquarters in order to carry out sliding stops and to turn in a graceful and efficient pirouette rather than lurching around slowly and awkwardly on the forelegs. Furphy demonstrates the difference when writing of the meeting between Collins and the stolid, bull-like British stockman who was riding a coarse cob,[400] which Furphy describes as a hippopotamus. In this case Collins daintily turned his horse, while the stockman 'pulled steadily on one rein and, so to speak, wore his ship of the plains round till we faced the cattle again.'[401] Thus, because pastoral breakers wasted little time on preliminaries and conventional mouthing, this stage of the process was particularly important. The emphasis here was on satisfactorily completing what would otherwise be seen as inadequate mouthing.

Although from the earliest days until the 1950s northern breakers occasionally broke in horses without yards, this practice has now almost certainly ceased. Barker deals with this method, especially its occurrence among drovers and teamsters in the 1930s. An unhandled horse was first captured by laying a noose on the ground and walking the horse plant over it. When the correct horse had placed a front hoof within the loop, the noose was pulled tight around its pastern. Inevitably the animal would plunge forward but, with two men holding the rope, it would be cartwheeled over as its front leg was pulled back underneath it. With this technique, most horses learnt within three heavy falls to stand still as they were approached and handled. In fact, Barker states that horses caught by this method were always the easiest to catch in mustering camps. Once the horse was on the ground one of the breakers would run forward and sit on its head, thus preventing it from regaining its feet (Plate 35, page 154). A head collar, lead rope and bridle were then fitted and the breaking-in process commenced. Because these horses were put into work as soon as they were ridden, 'extraordinarily quick progress' was made. When breaking in a horse by this method however, it was critically important that it did not break free with ropes and saddlery flapping behind. Should this occur the terrified animal was almost certain to gallop wildly, smashing heedlessly through the scrub.

> If it comes to a fence it will crash straight through it, [and] continue galloping perhaps for ten miles and in most cases die where it collapses. If it survives it would not be of any use, a nervous wreck and constitution ruined.[402]

The breaking-in of horses to pack-saddles took far less time and effort, for no mouthing and riding were involved. The animals had only to accept the pack-saddle girth strapped tightly around them, and not panic if the saddle-bags shifted or their contents rattled. No time at all was wasted on donkeys, 'We'd grab them, throw them down and chuck a pack on — break them in in one day.'[403] Almost without exception horses soon learnt that a greater clearance was now required in order to avoid trees and scrub. Mules, which are notoriously stubborn, were slower to comply. A claim is made that on one station in the Northern Territory during the 1960s, a group of mules being broken to pack-saddles simply ignored their altered width and persisted in pushing through narrow spaces, causing considerable damage to the saddles and bags. In a display of bush pragmatism three stout eighteen-inch lengths of wood were strapped to the bags, pointing out from each side and the top. High-tensile barbed wire was attached to the outside ends of these short poles, then taken forward and wrapped around the mules' necks. Consequently, every time these animals went too close to an obstruction they received a vicious jagging around their necks from the brutally sharp barbed wire.

Another technique used to facilitate the breaking and educating of pack animals was to give those difficult to catch the heaviest bags, so they were then willing to be caught and unloaded at the end of each day. Cole cured a mule of this trait by loading its bags with sand until it was carrying in excess of two hundredweight all day. He recounts that thereafter this animal was no trouble to catch. Sand was also used in a brutally pragmatic fashion to break the spirit of rogues. On one occasion in the early 1870s a team of Queensland drovers found they had been sold a horse which bucked extremely badly. It threw 'all the best riders of the party,' until no one would ride it. Accordingly, when they reached the Burdekin River crossing, near the junction of the Suttor River, the horse was run into a set of old yards, roped, thrown and saddled, and loaded with two saddle-bags filled with damp sand weighing approximately 450 pounds in total. A heavy greenhide rope was then wrapped tightly around the horse, saddle and bags. On regaining its feet the horse attempted to buck the bags off, but failed. It was then given several cuts with a stockwhip and driven with the mob for almost three miles across the heavy sand in the riverbed and the adjoining flood plain. That night:

the grey came quietly to be relieved of his pack. Anyone could ride him or do anything with him after that; but his spirit was utterly broken. He turned out an unmitigated slug. It was impossible to get a gallop out of him, and he was good for very little except carrying a pack.[404]

Finally, at the beginning of journeys, when pack animals were fresh and liable to bolt, with the risk of considerable damage to the saddles and equipment, stockmen employed a simple, inexpensive and highly efficient method of restraint. A hobble strap was applied to the pastern of one front leg and connected to the head collar by a side-line or whatever was available. By judiciously shortening this chain, the horse was effectively tied down and prevented from galloping or trotting. In fact it had difficulty walking and soon became tired. Thus, within a half mile or so it would be sufficiently settled to allow the chain or strap to be removed.

Draught Animals

While nearly all outback Australian properties were dependent on draught animals for the cartage of goods and supplies during the first 150 years of settlement, it should not be thought the industry was static. Indeed, the cartage industry in northern Australia changed continually in response to climatic conditions, the availability of different animals, the development of roads, the desire to increase load tonnages and the introduction of motor vehicles.

Traditionally bullocks were the mainstay of the industry. Not only were they cheaper and hardier than horses, but their higher water requirements were not necessarily a disadvantage in the better watered areas of Queensland. Horses came into their own, however, with the development of roads, for they could cover twenty miles in a day where bullocks would travel only twelve. In time bullocks were found primarily in the rougher and less accessible country. As Queensland pioneer Edward Palmer recalled in the late 1890s:

> The 'bullocky' was a great factor in the early days of settlement, where there were no roads and loading had to be dragged over mountains and through steep creeks and over all obstacles ... [but] The quicker-moving horse teams and the railways, are elbowing the bullock driver out into the never-never, where there are still opportunities for his special faculties, and it is not often that bullock teams, with their wood and iron yokes, and dusty, hairy drivers, are seen on any roads coming into railway stations.[405]

In the arid regions of the Northern Territory, the east Kimberleys and the Pilbara it was found that water could not always be guaranteed, and the feed was often inadequate to sustain horses and bullocks. This situation worsened as the more nutritious grasses and scrub were eaten out through overstocking, to the extent that in the Pilbara 'bullocks could barely live, let alone work.'[406] Supplementary feeding was out of the question, due to cost and distance. Thus, even though early photographs indicate that horses and bullocks were used initially in these regions, they were frequently replaced by mules and donkeys. In 1919 a *Northern Territory Times* correspondent reported that he had:

> met a team of forty donkeys conveying nine tons of payable loading for Ord River Station, and several other donkey teams consisting of between thirty-six to forty animals. I travelled close upon six hundred miles through that country, and met only one horse team — the Rosewood team. The Lissadale team is composed of half horses and half mules, and Argyle forty donkeys. Most of the other stations are using camels, and some are even breeding camels. There are eleven bullock camels working in a wagon in Wyndham, and the sight is one which it is almost worth while going that far to see.[407]

This somewhat incredulous reference to camels is interesting. In fact they were being used to pull wagons and buggies as early as the late 1800s, particularly in Western Australia's Eastern Goldfields, Murchison, Pilbara and east Kimberleys, for:

> In dry country camels were easily the most useful animals, donkeys came second, horses third, and bullocks a distant last.[408]

Little has been written on the use of camels in Australia. Initially they were often poorly received, though in large part this stemmed from ignorance and prejudice. Undeniably, some of the earliest cameleers found them dangerous and intensely frightening. Hermann Beckler, the botanical collector and medical officer on the Burke and Wills expedition, wrote:

> Matvala was the most feared animal, not only by the other camels, but also by us. He was a raging, to my mind crazy animal whom no one could handle except Beludsch. Even he had received a kick in the chest from Matvala in Melbourne, from the consequences of which he had never completely recovered. On this stretch it became my task, then, to guard this animal, that is, to keep it away from the others. Several times it attempted to escape, but each time I headed it another direction. Suddenly the camel became so furious that it turned on me, broke its hobbles at the first attempt and leapt at me with a terrifying bellow and with the wild fire of madness shining in its eyes ... It would certainly have reached and done for me had I not by chance found a bend in the vertically descending bank [of the river], down which I slid.[409]

H M Barker mentions that a number of the men who served in the Camel Corps in the Middle East during World War I came back with 'unpleasant recollections of camels.'[410] Among the specific charges levelled against them was their propensity to spit and bite. Barker rejects this, stating that:

> You often hear that camels kick and bite a lot. I have never been bitten by one, although they have made an open-mouthed dive at me and occasionally in this way they gave you a knock with their front teeth, which protrude forward. I have heard of people being bitten, but I have not seen it.[411]

He does admit, however, that when annoyed, camels would spit their cud into the teamster's face. Usually they did so when being harnessed or drenched, at a time when the unfortunate recipient was at close range.

> A big camel, as a means of defence, brings the cud up his throat, half a gallon I should say, and sends it out with about 30 lb pressure of compressed air behind it, rather a terrifying blast.[412]

In addition, while horses and bullocks were prone to wander at night, straying camels posed an even greater problem. On one occasion, when men were dying of scurvy and debilitation, Beckler awoke to find their camels:

> were gone. It was enough to drive one to despair! At a time when our lives could depend on half a day, all the animals had run off! ...
> In order to get a drink of water they had set off on a march of 50 miles though they could make only slow progress because of their hobbles. We soon found their tracks, but located them only several hours later ... marching pathetically along in their hobbles, while the majority must have broken their hobbles and were undoubtedly miles ahead ...[413]

As well, some camels worked with a pronounced rolling gait, and bumped into and antagonised those alongside them. Three rollers in a team were sufficient to cause the whole team to roll and teamsters usually discarded such animals. Furthermore, it was soon discovered that camels would die quickly from eating the foliage of ironwood trees. This markedly restricted their range of usefulness, for it precluded them from being used along the Murranji track and in other parts of the Northern Territory and the Kimberleys.

Nonetheless, camels had very considerable advantages. Foremost of these was their ability to withstand water deprivation. The explorer Ernest Giles, for instance, watered his camels on setting off from Ouldabinna in South Australia during his 1875 expedition. They were still in good shape when, ten days later, he found water at Boundary Dam on the Western Australian border. They then covered a further 323 miles over seventeen days, with only one small drink from supplies carried, before finding a plentiful source at Queen Victoria Springs, 130 miles east of the future mining town of Broad Arrow.

In arid regions a camel's ability to tolerate heavy work for sustained periods was also extremely valuable. Barker comments on this feature a number of times and outlines how some worked for three years nonstop. Even if they lost condition, it rarely seemed to affect their pulling capacity. Indeed, the limiting factor was the teamster, who usually required a break well before his camels.

In part, their resilience stemmed from the length of their necks, for they could obtain top feed beyond the reach of horses and bullocks. Furthermore, being ruminants, they could digest native herbage in areas where horses would rapidly lose condition. When Barker left Meekatharra during 1914, heading for his station in the Pilbara, the first three hundred miles of country were severely affected by drought. Although there was no feed available for his horse, which had to be fed on chaff, the camels 'always had mulga leaves and could mix them with leaves of a few other kinds of trees.'[414]

Camels were also capable of working for a longer time and covering greater distances each day than horses and bullocks. Whereas bullock teamsters usually only worked for six or seven hours and covered twelve to fifteen miles before camping and unyoking by mid-afternoon, camel teams frequently covered twenty miles laden or thirty miles unladen, and were still working at sunset. Furthermore, they worked in conditions that horses and especially bullocks would have found gruelling. Barker commonly worked in temperatures of 125 degrees Fahrenheit in the shade.

In order to capitalise on the advantages and overcome the disadvantages of the various animals and vehicles available, and to facilitate their everyday work, teamsters developed a variety of techniques, strategies and equipment. While bullock and horse teamsters to some extent drew on centuries of British and European experience in driving techniques, and harness and vehicle design and construction, and rapidly adjusted to the less forgiving Australian terrain, camel teamsters had no precedents from which to draw. They learnt as they proceeded, relying primarily on ingenuity and perseverance.

A case in point is the collars devised by camel teamsters after some experimentation. According to Barker, South Australian teams were initially held back by unsuitable collars, whereas his were made by a saddler in Perth who clearly understood the shape of a camel's shoulders. Cameleers also soon realised that, if left unsupported, collars would slip down onto the lowest part of a camel's neck. Thus the beast would be pulling from its neck rather than the muscular pads at either side of the neck on the shoulders. Accordingly, they devised an oblong leather harness, known as a spider, which fitted over the hump. Leather straps ran down from the front corners to either side of the collar and held it in the correct position for pulling, while a further two straps ran from the back corners to the trace chains and kept them at the correct height, so preventing the camel from getting its leg over them.[415]

As a result of such experimentation, within a remarkably

short time camel teamsters developed the skills necessary to pull huge weights and husband their stock. As Barker points out:

> Horses have been draught animals for centuries ... If a Queensland horse team and wagon was sold, then the buyer was sure of finding all the harness and other gear of the same pattern as he was used to, so had nothing to learn. Methods of driving the horses varied very little, and everyone knew the proper way to manage the work even if they did not always practise it. All this had evolved over a very long time ... Camels, on the other hand, were first used in harness in Australia only in the 1870s. At that time bullocks and horse teams were pulling really heavy loads — 10 to 20 tons — over soft, sandy or stony country, so camels were expected to do the same. It was found they could compete quite well.[416]

Probably the most important consideration for all teamsters was the selection of animals. Whereas taller, rangier bullocks were found to cope better than short-legged, heavy-set types, shorter, more thick-set camels were preferred for team work over the traditional longer, higher and leaner riding camel. Furthermore, in almost all cases castrated males were preferred to entires, for when camel bulls came into season they either fought continuously, trying to drive the other males away, or wandered.

> If one has a mixed herd, as we did, jealousy does not cease from the first day of the journey to the last. One of the male camels, whose sexual instincts are stirred up, wants to be alone with the females and so he chases away the other males with the most passionate fury, kicking and biting; that is, for as long as he feels himself superior to the others ... It became still worse when the male animal's passion reached a high level,

for he chased the entire herd before him for hours on end until he himself was tired out.[417]

Thus, while Afghan cameleers traditionally worked bull camels (though older or less assertive drivers worked cows), Australians mainly used geldings.[418] This was always a compromise, however, for bulls seemed to be hardier than the castrated animals. Those who took the risk and used bulls simply tied them down at night when they were in season. Indeed, Barker recollects how:

> A man from Meekatharra with five bull camels as a dray team tied them down for ten consecutive nights and worked them every day. They had all come on season at the same time and none would have fed if he had let them go.[419]

Whatever the animal, and irrespective of whether males, females or geldings were used, the initial handling, halter breaking and training could be difficult. Accordingly, some short cuts were developed. Ernest Henry broke his bullocks to harness by firstly lassoing them behind the horns and snigging them to a rail. Next, a hind leg was roped and they were drawn alongside the rails. A quiet bullock was then yoked to the wild animal for one or two days, before the pair were put into a team. The newcomer was usually tame and working well within a week. Considerable difficulties arose, however, when a whole new team had to be broken in and no experienced older animals were available. In the 1930s, after yoking his young bullocks in pairs and chaining the team together, former bullocky Arthur Cannon let them go in a paddock.

> There was an instant mix-up. The leaders started to walk away. But the second pair moved forward and tried to walk beside them, instead of behind. They tripped over the chain, got back-to-front with the

leaders and were pulled along backward. They got their yokes upside down. In fact, they managed to get into such awkward positions that more than once I felt like giving the job away. Eventually, I managed to fix a long link chain between the poler's start-ring and a big log. The latter, dragging behind, helped to keep the team straight. Also, it created a constant and wearisome burden that restrained any tendency to bolt. At least, bolting did not last so long.

The next week was a nightmare. Tangles, bolts, hang-ups — they were all part of breaking-in bullocks. I yoked them for four hours daily. That was enough for them — enough for me too.[420]

Ideally camels were handled as calves, for older animals were too strong to pull around from side to side, as was done with horses. Unhandled adults were simply chained around the neck and pulled behind a wagon for two or three days. By this means they learnt never to hang back and thereafter could be led with a head collar and lead rope. Pack-cameleers, who did not have access to a wagon, could only resort to tying the beast down, making a hole in its nostril and inserting an aluminium or wooden peg. The latter was preferred, for aluminium pegs often became uncomfortably hot during the day and irritated the camel. Once nose-pegged, a camel could easily be led.

In order to reduce the likelihood of stock wandering overnight, all teamsters learnt to ensure their animals were well watered and placed on good feed. Camel teamsters, mindful of their charges' proclivity to wander for many miles back along the day's route, also used fences to their advantage. Often they camped immediately past a fence, so the camels would only wander backwards and forwards along it during the night. Donkeys were the exception to the problem of straying, for they camped so well it was rarely necessary to hobble them. This was a mixed blessing for the teamsters, as a pack of braying donkeys nearby often made sleep almost impossible.

To simplify the everyday task of finding their stock, bullock teamsters frequently included two or three white or white-roan bullocks in their team, for they were readily visible for some miles in open country.[421] Dogs were also used, not only to help find and bring in the mob, but also to make them walk up to their respective positions among the yokes and chains and stand until they were yoked. Cannon, for instance, had a very good blue heeler bitch which enforced his orders with alacrity.

> A good heeler is wonderful to watch. Like a streak of greased lightning it races in and nips the heel of a reluctant bullock. How smartly it dodges down when the flying hoof comes towards it! And soon it goes in again when the hoof is back on the ground!
>
> Woe betide any beast that attempts to break away from the mob. Only young bullocks try. A sudden flash of hair, and the steer is lashing out at the thing that worries him. He soon returns to the team and behaves himself.

Unfortunately, she took a strychnine bait and died. Within a day the bullocks realised the dog was missing and 'began to play up.'[422]

Once the animals were found in the morning, the teamster would usually drive them in on horseback. Alternatively, camels were clipped by a cotton rope from their nose peg to the tail of the beast in front and driven in in single file. To prevent nose pegs being torn out and tails being broken, both of which mishaps were extremely painful, a short length of string was inserted in the line. This would snap before serious injury could occur.

Obtaining the maximum effort from teams, and keeping them fit and sound while working under arduous conditions for extended periods, demanded an extraordinarily wide range of skills. Certainly this was not a job for what Furphy called 'Apostles of brute-force,' a term he used when explaining why the relentlessly hardworking bullocky, Priestley, never got ahead.

> Some carriers never learn the great lesson, that to everything there is a time and a season — a time for work, and a time for repose — hence you find the industrious man's inveterately leg-weary set of frames [bullocks] in hopeless competition with the judiciously lazy man's string of daisies. The contrast is sickening ... But the Scotch-navigator can't see it. He is too furiously busy for eighteen hours out of the twenty-four to notice that, even in the most literal sense, loafing has a more intimate connection with bread-winning than working can possible have.[423]

Foremost of these skills was the setting-up of the team. The first consideration was to keep it as close as possible to the dray or wagon, for the further the animals stretched in front the less efficient they were. Donkeys, which are not only short in length but were also often hitched four abreast, could pull disproportionately heavy loads for their size relative to bullocks, which were always yoked in pairs and well strung out. In 1942, for example, an *Electra* plane crash-landed on Napier Downs in the Kimberleys. After it was repaired with parts flown in by a *Cessna*, the plane had to be dragged from the swampy ground where it had landed, and was now bogged, to a newly cleared airstrip 250 yards away. Local station manager Ned Delouer suggested using a donkey team and twenty-seven were harnessed three abreast.

> The first attempt ... failed as soon as it started. The animals sank up to their bellies and had to be dug out. Harnesses became snarled as animals bogged and others fell over them.[424]

According to Delouer's daughter, several of the team had not

Plate 53: Ned Delouer and his press-ganged donkey team, Napier Downs, 1942.

been broken to harness and many of the ill-handled animals kicked frenziedly at their harness as they bogged, dragging others down. Delouer and his assistants were nonplussed and eventually order was restored. On the second attempt:

> the plane began to move steadily until mud clogged the landing wheels and they began to slide rather than roll. Delouer was not worried; it was moving and he aimed to keep it moving.
>
> At one point it crashed into a creek but by now the donkeys were on firm ground and able to pull it out [Plate 53].[425]

Donkeys exhibited similar tenacity in road-making operations at Broome during the 1920s. At this time the Broome

Shire used one of the earliest available four-wheel-drive trucks for dragging sections of railway line to form roads. Unfortunately the vehicle had solid rubber tyres and almost invariably became bogged when crossing creek-beds or patches of loose sand. As a result it was generally followed by a team of donkeys, which dragged it clear after each ignominious failure. Consequently local wags claimed that the FWD insignia stamped on the bonnet did not stand for 'four-wheel drive', but 'f*cked without donkeys'!

Configurations varied, though, depending on the type of vehicle, the type of animal available and local conditions. When working in very dense scrub or along a narrow track, teamsters were often forced to harness their team only two abreast. Yet too long a team could be a significant cause of accidents. If the leaders were too far in front and, in effect, pulling from the side, a wagon could be pulled over the edge while winding down hills in a zig-zag fashion. Accordingly, Barker soon learnt to unhitch two-thirds of his camel team when descending the reputed two-thousand foot decline from the Hamersley plateau to the Roebourne coastal plain.

Placement of the leaders was also important. Because almost all draught animals pulling loads were not driven with reins, but steered with the universal commands 'whoa', 'whoa-back' and 'gee-off' (that is, stop, turn right and turn left), when working in thick scrub Barker often placed his most obedient camel in front by itself. In this situation, without it having two other camels to push against as it turned to the off side, far greater control was possible.

Whatever the configuration used, teamsters also had to take into consideration the anatomical restrictions of their stock. For instance, while bringing heavy loads down steep hills posed problems for all animals, camels were especially badly affected. Whereas the two horses harnessed immediately to the cart — the shafters — learnt to sit back on the heavy leather strap which passed from shaft to shaft just below their hindquarters, in order to reduce the strain on their forelegs, camels could not do this.

Understanding how and when to use whips was extremely important in keeping individual animals working honestly. A bullocky's whip consisted of a wooden handle six to seven feet long and an eight-foot plaited lash. Generally he used the whip to indicate his intentions, but would flog the team with it in heavy pulling. The driver almost always walked on the near side of the team and, with the length of the whip plus his arm, had an effective range of sixteen to eighteen feet forward and backward. As each yoke of bullocks spanned eight feet, seven yokes (fourteen bullocks) covered fifty-six feet, so bullockies had to keep moving backwards and forwards along the team. Usually the better, more honest, pullers were put on the offside, so the driver had easier access to the less reliable animals. According to Cannon, an assistant sometimes worked on the offside, thus the term 'offsider'. In the few situations where a long-handled whip could not be used, such as when the driver rode his horse alongside the team on a long straight stretch of good pulling, or when working in heavy scrub, shorter, flexible-handled bullwhips were used.[426] Camel-drivers' whips were slightly shorter than bullockies' whips and, because of the higher-set chains, they could only be used on the middle camels' hind legs and on the nearside camels' lower body.

Teamsters soon became aware that different animals responded differently to whipping. Horses and camels, for instance, were more sensitive and reacted badly to excessive use of the whip. Curiously, Barker found that his camels took little notice of being hit on the body, but responded well to being hit on the legs or underbelly. As overuse of whips diminished their effectiveness, teamsters learnt to use them judiciously. It was essential that only those animals which were not working honestly be hit. Campbell describes how on one occasion he suddenly realised some of his donkeys were deceiving him.

> They made out they were doing a magnificent job, but were really just keeping their chains tight. Their small hoofs barely dented the ground. As a result of this

discovery I uncoiled my beautiful whip and let the slackers have it. The honest donkeys did not mind and did not change their pace. The slackers leapt as the lash flashed out and left three stripes on their backsides.[427]

For this reason Barker was critical of Aboriginal teamsters who, he claimed, frequently spoiled teams through excessively and indiscriminately hitting even the good pullers.[428]

In order to reduce reliance on whips, teamsters named every animal in the team so that, when called upon, an individual would respond with alacrity. Although bullockies traditionally used names ending in 'ie' or 'y', such as Bluey, Strawberry and Darkie, names ideally were of one syllable and distinct from all other names, so there was no confusion when a teamster called on an animal to respond. Camels and horses were also usually fitted with winkers to make them concentrate on their work and pull more truly.

Teamsters had also to harness, so to speak, the different pulling styles of the various animals. Horses, for example, could often get a wagon out of a bog by snatching forward into their collars, or up a steep creek bank by taking advantage of a downhill lead-up. By contrast, bullocks and camels firstly took the strain and then, at a signal from the teamster, usually a whistle, perceptibly lowered themselves before exerting the maximum effort. Once under way, they could sustain the maximum pull for longer than horses. Camels were especially noted for this.

After a period of sustained hard pulling, teamsters stopped and rested their teams for some minutes. This was essential to allow them to recuperate. Without such pauses animals could die of heart attack, stroke or overheating, as both their heart rate and blood pressure were seriously elevated by such exertion. Prospector A J Macgeorge of Kalgoorlie recalls that horses frequently died when pulling in the hot sun through a twenty-mile long heavy sand plain east of Southern Cross during the 1890s. Care had to be taken, however, that the team was well clear of a patch of heavy going before stopping, lest they develop the habit of stopping immediately they passed a heavy patch.

Teamsters also learnt a number of small skills which, collectively, helped maintain their stock. As early as 1830 they were aware that bullocks pull best when warm. Accordingly, Mitchell attempted to place his campsites in locations from which his stock would have relatively easy country to cover while they warmed up in the mornings. Eyre also favoured an easy beginning to the day.

> I generally preferred, if practicable, to lengthen the stage a little in the vicinity of watercourses or hills, in order to get the worst of the road over whilst the horses worked together and were warm, rather than leave a difficult country to be passed over the first thing in the morning, when, for want of exercise, the teams are chill and stiff, and require to be stimulated before they will work well in unison.[429]

Similarly, Barker believed his camels endured their work so well because he left them sitting in their harness for an hour or more at the end of each day before releasing them.

> They enjoyed the spell and expected it. They chewed the cud for a while and one by one stretched their necks out with chins on the ground and slept. After I had removed their harness, they usually continued sleeping until [their leader] Diamond got up, shook himself and rang the big bell; then they would all move off and begin feeding and would be found next morning nearer by far than if I had let them go as soon as camp was reached.
>
> Mohammed Hassan, Cumming's head camel man, advised me to do this. All the Afghans did it and most white men scoffed at the idea, but I am positive it was sound. Hassan said it had been done whenever possible for thousands of years.[430]

In this light, when faced with the choice of halting for lunch, bullock and horse teamsters usually went straight through to their camp before stopping, thus allowing their stock to be unloaded, watered and put out to graze some hours earlier. By contrast camel teamsters, who usually worked longer hours, often stopped and spelled their camels at lunchtime, although finding a suitable resting-place was not always easy in very hot areas. Pockets of deep sand would scald the camels' sensitive underbellies, so drovers looked for a smooth, hard claypan where the camels could lie down with their legs underneath them and chew their cud.

The meticulous planning and husbandry necessary to overcome dry stages, which inexorably increased in length as the dry season progressed, were also gradually developed. While bullocks were especially affected, all teamsters feared being 'locked in' at a waterhole and unable to advance or retreat until the wet season arrived. In her vividly evocative style, Jeannie Gunn describes the two Macs' technique.

> With well-nursed bullocks, and a full complement of them — the 'Macs' had twenty-two per wagon for their dry stages — a 'thirty-five-mile dry' can be 'rushed'. The waggoners getting under way by three o'clock one afternoon, travelling all night with a spell or two for the bullocks by the way, and 'punching' them into water within twenty-four hours.
>
> 'Getting over a fifty-mile dry' is, however, a more complicated business, and suggests a treadmill. The waggons are 'pulled out' ten miles in the late afternoon, the bullocks unyoked and brought back to the water, spelled most of the next day, given a last drink and travelled back to the waiting wagons by sun-down; yoked up and travelled on all that night and part of the next day; once more unyoked at the end of the forty miles of the stage; taken forward to the next water, and spelled and nursed upon again at

Plate 54: Delivering pumping plant to a bore site, Wave Hill station, 9 September 1921.

this water for a day or two; travelled back to the waggons, and again yoked up, and finally brought forward in the night with the loads to the water.

Fifty mile dry with loaded waggons being the limit for mortal bullocks, the Government breaks the 'seventy-five' with a 'drink' sent out in tanks on one of the telegraph station waggons. The stage thus broken into a 'thirty-five mile dry,' with another of forty on top of that, becomes complicated to giddiness in its backings, and fillings, and goings and comings, and returnings.[431]

The design and maintenance of vehicles also underwent considerable experimentation and refinement. An extensive array of single- and double-axle vehicles, ranging from the stolidly pragmatic wagon shown in Plate 54 to elegantly designed and constructed, beautifully painted, leather- or steel-sprung, light-weight sulkies, conveyed men, women and children millions of miles in complete safety. Nevertheless, all these vehicles suffered from inherent shortcomings. For one thing, many were ill-suited to off-road conditions, as a number of commentators humorously pointed out. The newly married Gunn, for instance, recounts how, when being driven from Katherine to Elsey station, they:

> set off in great style — across country apparently — missing trees by a hair's-breadth, and bumping over the ant-hills, boulders, and broken boughs that lay half-hidden in the long grass.
>
> After being nearly bumped out of the buck-board several times, I asked if there wasn't any track anywhere, and Mac once again exploded with astonishment.
>
> 'We're on the track,' he shouted. 'Good heavens, do you mean to say you can't see it on ahead there?' and he pointed towards what looked like thickly timbered country, plentifully strewn with further boulders and boughs and ant-hills; and as I shook my head, he shrugged his shoulders hopelessly. 'And we're on the main transcontinental route from Adelaide to Port Darwin,' he said.[432]

Even when roads were available, they were usually of such a poor standard that passengers still suffered appallingly. When travelling in a Cobb and Co. coach in western New South Wales during 1880, French observer Edmond La Meslee savoured the scenery for only a short distance from the township before:

> a violent jolt announced that we had left the made road altogether.
>
> … Suddenly, catapulted from my seat, I was hurled into the arms of the traveller sitting opposite: the Count soared up to the roof where his top-hat was flattened like a pancake: my neighbour disappeared under the

> seat as he threw the child into its mother's lap. This lady, to fortify herself for the journey, was carrying two great flagons of eau-de-cologne which she had been applying to each nostril alternately. Flying from her hands, they broke on the head of the second passenger and we were all rendered embarrassingly fragrant for the rest of the journey. The unhappy man, whose head had been anointed, swore like one possessed.
>
> It took us some minutes to recover from the shock, but we had thirty-six hours to travel in this fashion. To counter the terrifying lurches that followed each other in rapid succession, we wedged ourselves into position as well as we could.
>
> Nothing can give any real conception of a two-hundred mile journey through the bush in a virgin wilderness, without roads worthy of the name, in impossible coaches slung on leather springs (steel ones won't stand up to it), and pulled at full speed by four spirited horses. Never stopping for a moment we brushed past overhanging branches, tree-trunks and yawning ruts at the gallop, while the infernal machine threatened to overturn every moment.[433]

Discomfort aside, horse-drawn carriages were intrinsically dangerous. On one occasion La Meslee and his travelling partners were forced to replace two of their horses, which had been poisoned by cyanide tailings at a mine site. One of the substitutes was a black horse named Zanzibar, which was 'well known in the district as a good racehorse, but as being very vicious if put into harness. Of the last report we were ignorant.' Consequently, when towards evening the driver cracked the whip over the horses' heads, Zanzibar leapt straight to a gallop and took the others with him.

> Before Mr Barton had time to check them, all four, with outstretched necks and flaring nostrils, were

hurtling us along at the speed of an express train.

No human power could save us.

We were at the mercy of the four animals, intoxicated by their own speed and convinced that they were in the midst of an exhilarating contest on the race-course.

It was impossible even to try to stop them!

Barton, the driver, worked coolly on the team, but could not prevent the coach swaying wider and wider as the track swerved around trees. All too soon a huge tree loomed, and La Meslee and the others sat transfixed until, with:

> a terrible crackling, the sudden noise of breaking branches, the flying vehicle abruptly brought to a halt ... and we realised that we had, literally, been projected into a tree.
>
> Having seen at a glance the absolute impossibility of driving round the gum tree which barred our way, Mr Barton had steered his bolting team into a mulga thicket on the opposite side of the track ... There we were safe and sound, staring stupidly at each other and asking ourselves if the whole thing had not been a dream, and whether we were indeed still alive.
>
> The sight that met our eyes quickly brought us back into the realm of reality.
>
> Chaos was complete.
>
> Horses, carriage and passengers lay in a confused heap. The smash had been such that the lead-horses, suddenly stopped in their mad career, had crashed to the ground. Carried forward by their own momentum, the wheelers had been thrown on top of them; the dog-cart had ridden up crazily over all and we were roosting, without a scratch, among the branches of a big mulga.
>
> We were soon on the ground and ... righted our conveyance and got the horses to their feet. They were

Plate 55: Bush wagonette built at the blacksmith's shop, Manbulloo station, by wheelwright Bray and blacksmith Gaynor, 25 June 1921.

trembling with nervous terror but, by some inexplicable stroke of luck, did not seem to have suffered any injury. It was quite otherwise with our vehicle: the pole had been shattered in three places and the fore-carriage buckled. But thanks to our ample supply of ropes and leather thongs, without which no-one undertakes a bush journey, the damage was temporarily repaired within a quarter of an hour, two of the horses were re-harnessed, and we limped off towards the Shear-legs Hotel.[434]

Women passengers had even more problems to contend with. Four-wheeled buggies, commonly known as traps, had a step between the wheels, and the women had to gather their long skirts and lift them free while they raised their foot to the step. Given the height of some wheels, such as those shown in Plate 55, this frequently required considerable effort. Moreover, even should they manage to reach the step and pull themselves up before the frequently excited horses moved forward, they risked having their clothes smeared with mud and dirt, or catching them on the foot-brake. Once the women were aboard, babies and small children had to be handed up, whereupon:

> It was a common thing for horses to bolt with a buggy, perhaps leave the road, gallop across country and smash the buggy. If that happened someone was sure to get hurt, in some cases having bones broken. Babies proved to be almost indestructible. Adults would be thrown out and injured while the baby, when picked up, was found to be none the worse.
> To climb up into a sulky or two-wheeler was equally dangerous ...

Thus it is hardly surprising that 'women threw their influential weight into favour of motor cars'[435] and horse-drawn personal transport was quickly replaced. The demise of the cartage industry occurred slightly later.

In the early days most teamsters used single-axle, two-wheeled drays for cartage purposes. These had shafts or a single pole protruding forward and were pulled by teams ranging in size from one to ten animals. Drays had severe deficiencies, for their single axle meant they could tip up 'going out of a steep creek with a load on, and going down would bear on the polers fit to break their necks.'[436] Thus it is eminently understandable that Edward Palmer described them as a 'fearful kind of vehicle.' Certainly teamsters soon realised the limitations of these 'bullock-killing' vehicles, and consequently introduced double-axle, four-wheeled wagons, which were pulled by teams almost twice as large as dray teams. The design principles for these vehicles were essentially the same throughout Australia. Wagon wheels, for instance, were traditionally set so they leant outwards slightly at the top. When loaded, the axle would bend slightly and thus bring the wheels to the vertical. Without this offset the wagon would have been heavier to pull.

Some modifications were necessary for wagons working in the more arid northern regions. The main change was that wheels had to be made from inland timber, which could tolerate hot and dry conditions. Barker was advised not to buy a wagon made on the coast if he intended taking it to the Murchison or Pilbara, for 'every single wagon made on the coast suffered through shrinkage of the timber in the wheels.'[437] After making inquiries he was able to purchase a secondhand wagon made by Toohey and Thornton of Nannine, which could carry twenty tons. There was no question of these wheels shrinking, for the original owner of the wagon had cut the timber within the district and left it stacked for some years. During this time he had turned each piece of wood several times a year, to allow it to season thoroughly. Furthermore:

> The timber for the wheels was york gum, a wonderful wood for wagons. Wheelwrights said there was nothing to equal york gum, but it was tough to work. The timber for the felloes, the 6 inch by 6 inch rim of the wheel, was selected for its natural curves, curves which were the same as the wheels they were to form. This was much better than cutting curved felloes from straight logs with a bandsaw, irrespective of the way the grain ran.[438]

Barker obtained a further two decades' work from this wagon, without any problems with the wheels, before re-selling it. This was as much a tribute to his care as to the wheelwright's skill, for he jacked the wheels up every two years and spun them in a trough of heated linseed oil. Not only did this preserve the wood, but it also prevented it from becoming waterlogged and swollen in the wet season. Toohey and Thornton had also helped considerably by forming the ends of the spokes with square shoulders rather than the cheaper bevelled shoulders, thus preventing the wheels from being pulled out of round.

On one occasion, also, he replaced the outer steel tyres. Wheelwrighting was an exacting trade. For one thing, the tyre had to fit tightly, but not too tightly, on the wheel. A proven formula was used to calculate the amount by which the circumference of the tyre had to be smaller than that of the wheel. Having been made to size, it was heated until it glowed cherry-red in the dark, at which point it had expanded sufficiently to be dropped over the wooden wheel. Once on the wheel, it was rapidly doused in water before it could do any damage. As the tyre cooled it contracted and pulled tighter and tighter on the wheel, until it began to stretch. With the correct formula, the tyre then had a firm grip on the wheel, but was not so tight as to damage it.

While the process sounds relatively straightforward, it was a dangerous, exacting and demanding task which required large fires, lifting cranes and dousing ponds. A tyre could weigh up to a quarter of a ton, and should it not be dropped on the wheel correctly, the wheelwrights were in terrible trouble. Frequently all hands would:

> lose their heads, some trying to lift it off, and others chucking water on when they see the woodwork burning. The air becomes thick with the smell of burning paint, smoke and steam, not to mention oaths. Someone is sure to get burnt with hot steam and the height of unpleasantness prevails. When tyres and tempers have cooled down, there she is, a ruined wagon wheel, tyre half on and half off, quite immovable.[439]

Yet despite this immense amount of carefully refined knowledge, motor vehicles made very rapid inroads into the use of draught animals and the end was swift. In the early 1900s, for instance, Herbert Hoover, later to be President of the United States of America, was the Australian manager of Bewick Moreing and Company's eastern and Murchison goldmining operations. Each week he spent long, arduous hours covering four hundred miles by horse and buggy while inspecting mine sites. When he purchased a Panhard car from France he was able to cover an 'unheard of 125 miles in a single day.'[440] Similar feats were achieved as the century progressed, and it is hardly surprising that one day in 1923, when Barker was carting asbestos from Nullagine to Marble Bar, a distance of seventy miles, he was overtaken by a truck. The owner of this first vehicle in the district had commenced carting from the same mine and he delivered three loads in the time Barker took to reach Marble Bar. In other words, the truck was working more than six times as quickly. Sadly, Barker was not initially perturbed, as he believed the truck would not be competitive, and he was most surprised when:

> the mine manager sent word to me not to come back for more asbestos ... Worse was to follow, for every year from then on came more and more motor lorries and fewer and fewer camel and donkey teams could find work in the district ... My recollection of that motor revolution is that we were in a state of bewilderment while it was going on. It seemed too good to be true, and there surely must be a catch to it. Camels and horses had done that work from the dawn of history, thousands of years, and that they were to be discarded for good in our short time was too big a fact to accept without question or doubt. What was quite unexpected was the way even the most conservative of station owners took to motors readily. Some had defied every modern invention, such as machine shearing, all their lives; yet they took to motors straight away. Looking back at it all now after thirty odd years, it can be said that the machine age, or rather the motor transport age, fairly burst upon us. In ten years, it was all over.[441]

Safety and the Use of Firearms

According to historian Henry Reynolds, close to two and a half thousand white settlers and twenty thousand Aborigines were killed in conflict during the occupation and settlement of Australia.[442] So fierce was the struggle at times that in one year during the late 1850s, in the Taroom region of southern Queensland, twenty whites were killed from a population of approximately 180.[443]

Certainly, for white and black alike, violence and death were facts of life. As Reynolds points out:

> Almost every district settled during the nineteenth century had a history of conflict between local clans and encroaching settlers. Many of the Europeans who lived through the time of confrontation were quite realistic about the human cost of colonisation. A small town pioneer wrote in 1869 that his community 'had its foundations cemented in blood.'[444]

The factors underlying this combative period of our early history warrant brief reiteration. Easily the most important were the introduction of livestock and the dispossession of Aborigines. What many contemporary urban critics of this period of pastoral history fail or perhaps refuse to understand is that, from the moment livestock were driven into a new region, the tribal owners' traditional lifestyle was inexorably doomed. Within days, or at most weeks, stock began to foul and empty waterholes and compete for grazing. As a consequence native game became depleted, or moved further out to undisturbed areas. The Aboriginal owners, being constrained within their traditional lands, did not have that option. The results were almost inevitable. Reynolds cites missionary Francis Tuckfield, who in the early 1840s commented that there was very little food left in the bush in the Western District of Victoria. At much the same time Assistant-Protector E S Parker observed that:

> the Natives are now in a much worse condition and present a far less robust appearance than when [we] arrived — and that it is [our] decided conviction, that they must occasionally suffer great privations from their altered and often emaciated appearance.[445]

Accordingly, Aborigines were compelled to kill pastoral livestock. These depredations were almost inevitably met with force by the new settlers, many of whom had borrowed heavily in an attempt to stock their properties and were ill-placed to afford any losses. Reprisal led to counter-reprisal, the pillaging and despoiling of settlers' supplies, the slaughter of large numbers of livestock and the eventual indiscriminate murder of men, women and children. Tales of these interactions often preceded settlement, so that increasingly both Aborigines and pioneers in newly settled areas were frightened, intolerant and potentially violent.[446]

This cycle of violence was exacerbated by incompatible cultural values. Misunderstandings, especially over the

Aborigines' habit of carrying spears long after they were allowed to return to settled areas,[447] were potent sources of mischief. Early settlers were also initially appalled by the Aborigines' apparent indifference to, or at least casual acceptance of, violence and murder. West Kimberley pioneer Bob Thompson, for instance, told Ion Idriess in the 1930s that his station hands would murder him except that the annual mounted-police patrol would eventually learn of the atrocity and that after his death they would receive no more tobacco![448] Certainly, Aborigines came to be seen in a very unfavourable light. Alfred Searcy, for one, believed that 'The nigger, even the half-civilised one, is an uncertain animal, and always possesses the brute desire to kill.'[449]

Traditional notions of revenge were also initially seen as abhorrent, these being premised basically on the principle of an eye for an eye.[450] Beyond that, tribal Aborigines were not tightly constrained. Indeed, any person could be commensurately injured or murdered in retaliation for an assault, so long as they came from the same tribe or area as the transgressor. The seemingly non-tribal whites presented a conundrum which the Aborigines appear to have overcome by accepting that any whites, whether implicated or not, could be murdered in response to injuries inflicted by white settlers. Lieutenant Barrallier commented on this striking cultural difference as early as his Blue Mountains exploration of 1802. As he correctly pointed out:

> It is not of any advantage but, on the contrary, it is very dangerous to offer any insult to the natives. They avenge themselves of it sooner or later and the first white man they meet without means of defence becomes their victim. They make use of the most cruel tortures on the one they can catch, whoever he might be, without troubling in the least about enquiring whether he belonged or not to the party who ill-treated them.[451]

Unfortunately, this crucial information did not become common knowledge until some time after this. Consequently, when tribal Aborigines commenced revenge killings against Europeans, they (quite reasonably by their standards) frequently murdered innocent parties, whereupon the white settlers, not surprisingly, were incensed and so had no compunction about reciprocating. The difference in intent was, however, that the outnumbered settlers were bent on radical change and total subjugation. They met violence with greater violence and, in the short term, further aggravated the cycle of murder and destruction. William Fraser, one of the two surviving members of the Fraser family massacred at Hornet Bank near Taroom in Queensland, is reputed to have murdered in excess of a hundred Aborigines in retaliation.[452]

Settlers also became extremely mistrustful and increasingly defensive in reaction to Aboriginal duplicity. This phenomenon appears to have been relatively consistent throughout Australia, for it was reported by several early explorers including Sturt and Mitchell. Indeed Mitchell, notwithstanding his reliance on and friendship with these people, was the most outspoken of the major explorers in his appraisal of their character.

> Treachery and cunning are inherent in the breast of every savage. I question, indeed, if they are not considered by them as cardinal virtues ...[453]

Probably the most frequently cited example of Aboriginal duplicity is the incident at Hornet Bank. In this instance the Aboriginal odd-job man Baulie, who was originally from New South Wales, reputedly conspired with the local Jiman people by clubbing the watch-dogs to death to prevent them warning the Frasers of an impending attack. He was also present at the sexual assault and murder of the female family members.[454] Billy Ward, or Brigalow Bill as he was known, the original settler on Humbert River, is also said to have died as a result of such behaviour. According to Charlie Schultz:

The day Brigalow Bill was killed he was up at a yard near his house handling a young colt. He had a lubra called Lulu and she was up at the yard giving him a hand. She noticed that for the first time he didn't have his revolver on his belt — he'd left it down at the camp. So she told him, 'I go down, boil 'em up cuppa tea.' Brigalow said, 'All right. I'll finish handling this colt.' When Lulu came down she got his revolver from under his pillow, put it in a bucket and took it down to the waterhole below the house. There was a big camp of bush blacks across the river from what I heard, and she told them she had Brigalow's gun. That's when they came up to kill him.

About half an hour later Brigalow finished with the colt and started down to the hut. As he was going down he saw three Aboriginals walking towards him — Gordon, Maroun, and I forget the other fella's name. They had spears, and spread out as they walked. He knew then that he was in for trouble. He stopped till they got closer, then made a rush for it. One spear got him in the arm, but he pulled it out and threw it straight back at the nearest blackfella ... Then Brigalow ran on down to his camp and dived his hand in under his pillow to get his revolver — it was gone.

The blacks poked at him there in his humpy for quite a while and then made as if to leave. After about an hour Brigalow couldn't see anyone around, so he thought he'd make a run for it. He wrapped a bridle around his shoulder and neck, and ran straight towards the yard to try to get a horse. He didn't get far. As he made a run for it, Gordon stood straight out in front of him and let fly with a shovel-nosed spear — got him fair in the stomach. Before they finished him off the women came and urinated on his face. Then they took what they wanted from his house and after that most of the blacks camped across the river took off up the Humbert and scattered to the seven winds.[455]

Similarly, in 1895 two teamsters, John Mulligan and J Ligar, were attacked at Jasper Creek Gorge, fifty miles inland from the Victoria River Downs depot. After supper their three Queensland Aboriginal stockmen visited a nearby Aboriginal camp, taking their rifles as was their usual custom. About eight o'clock, while the teamsters were:

> standing together near the wagons, they were suddenly surprised by a shower of spears. One of the spears (a murderous weapon made of the blade of a stolen sheep-shears) struck Mulligan in the thigh, whilst Ligar was wounded in the back with a stone-headed spear, which penetrated the lung, and also with a glass-headed spear, entering the right side of the face and penetrating to the left cheek bone. Mulligan was the first to recover from the shock of the unlooked for attack, and succeeded in keeping the black marauders at bay with his rifle, Ligar being temporarily rendered incapable of lending any assistance by the rush of blood from the frightful wound in his face. The men remained in this terrible position throughout the remainder of the night, the natives renewing the attack again and again; during the fighting the voices of Mulligan's black-boys could be distinctly heard inciting the other natives. When morning broke Mulligan assisted Ligar to the top of one of the wagons to throw down some bags of flour with which to form a barricade. Whilst Ligar was thus employed one of Mulligan's boys, known as 'Major', fired at him with a rifle, showing unmistakably that the renegade Queensland boys had leagued with the local natives ... The attack was kept up off and on for

three days, the blacks only being kept from rushing the camp by a wholesome fear of the white men's firearms. Ligar was by this time becoming very weak from loss of blood and want of rest, and Mulligan decided that they should try to make Auvergne Station, the nearest point at which assistance was to be obtained.[456]

Accordingly, after creeping out and catching two draught-horses, they rode eighty miles to safety. To whites, the most disconcerting feature of this event was that, over the two days during which these men were attacked, shots were fired by their native assistants, who had joined the assailants.

Behaviour of this kind is hardly surprising, however, given that station Aborigines, who in many instances had only recently come in, had conflicting loyalties. Often they were duty bound to assist tribal Aborigines bent on retribution. Indeed, as local frontiers were in a constant state of flux for decades, with Aborigines moving in and out of pastoral service, their loyalties may have been wholly with their kin. Consequently it is not surprising that they frequently offered support to tribal and non-tribal Aborigines who refused to be coerced into pastoral service.[457]

Problems also arose from differing notions of reciprocity and obligation. The settlers' failure to meet their obligations, especially those arising from sexual favours granted by Aboriginal women, was a particularly potent source of ill-feeling and violence. A further important factor was the perceived despoiling by whites of critically significant religious sites, and their refusal to allow Aborigines to visit areas for vitally important ceremonies and observances.[458] In short, the forced interaction of these highly disparate cultures led to acute misunderstandings and inexorably worsening violence.

Although Reynolds and other revisionist historians of the past two decades have dealt with this question comprehensively, their work is not without shortcomings. While vigorous in their analyses of Aboriginal trauma, they have failed to convey an equal understanding of the intense psychological sufferings of the isolated, grossly outnumbered and often ill-equipped early white settlers. Unarguably, Aboriginal Australians initially enjoyed many advantages in this conflict. Foremost of these were their bush skills, which had been developed and refined for up to forty thousand years, and their intimate knowledge of tribal terrain. Accordingly, they could reconnoitre and close in undetected, then slip away easily, quickly and safely. Moreover, the scrubby, uncleared topography of much of Australia meant conventional military tactics could rarely be applied against Aborigines. Whereas the vast American prairies frequently permitted the United States Army to employ cavalry and even field artillery, the Australian terrain generally precluded such operations. Another factor was that the general paucity of native foods, which tended to preclude extended large-scale gatherings, largely determined the Aborigines' extremely effective small-scale, fluid style of fighting. Clearly, the Aborigines' tactics were those of guerilla warfare, and it is not surprising that their considerable superiority provoked anxiety and fear.

> The manner in which the blacks fought struck terror into the hearts of the settlers. No one was safe. At any time, day or night, a party of blacks might sneak up and, with wild yells, spear men, women, and children, old or young, without warning ... If they could not effect a surprise they withdrew and waited ... The whites all carried arms when travelling, and even while working about their homes. Shepherds and other workmen went in pairs. There was no safety anywhere outside the cleared lands round the larger towns.[459]

Not surprisingly, early settlers were always watching and waiting, fearfully anticipating surprise attacks by Aborigines. Schultz's experience on the highly vulnerable Humbert River station in the late 1920s exemplifies this unceasing stress. On this occasion, having lived alone for over twelve months, apart from his station Aborigines, he was awakened early on a pitch-dark morning.

> I heard the dogs barking. I was always a very light sleeper those days and I woke. I heard one dog bark and then another growling. A half-grown cattle dog was chained up right near my door and I heard him bark when the other dogs weren't barking. I thought 'That's funny?' and sat up in my bed. The young dog barked again.
>
> There were still wild blacks in the bush then, so I always camped with a revolver under my pillow and a rifle alongside of me — always loaded of course. I scrambled out from under my bush net and grabbed the rifle. I had the sense to stand to one side instead of in front of both doors, so no one could pick me off. I knew they couldn't do anything from the sides because the walls were solid timber.
>
> I sang out: 'Who's there?' No answer. I waited a while and the dog started growling again. I could hear him moving on his chain. 'That bugger can see something and it's not likely to be a dingo,' I thought. I cocked the rifle. I stood right near the door then in case anyone was going to come through. A thousand thoughts flashed through my mind: 'Is a mob of blacks going to rush me? Will I shoot on sight?' If it was blacks I knew they couldn't give much trouble outside, but they could at the doors, so I just stood there with the rifle cocked and I thought, 'Well, I'll get the first bloke anyway.'
>
> I called out: 'If you don't answer I'll shoot!' Still no answer. I was getting more stirred up all the bloody time, and next minute the dog sat back on the chain as though somebody was trying to wallop him with a stick. With that, I jumped outside quick smart with the

rifle ready. I was looking a dozen places at once, trying to see who or what was there, but about all I could see in the dark was the bloody dog sitting back on the chain. Then the dog started yelping like hell and the next bloody minute something hit me on the leg. Oh God — I nearly went through the sky!

It was a bloody chook! I had some fowls that used to roost in a tree alongside the house, and one fell out of the tree apparently. They get blind at night and it blundered down towards the house. It was heading up to this half-grown cattle dog, and the dog was taking ever-increasing fright. And then it hit me on the leg. By gees, I tell you what, that gave me a bloody fright. I went inside and I sat on my bloody bed, and I thought, 'Is it worth it?' I'd had it. I was twenty or twenty-one and I cried that night — I'd really had it. I thought, 'Bugger this, life's not worth it. I'm gettin' out of this place tomorrow. I'll ride over to VRD tomorrow and I'll leave my bloody horses there and I'll get out somehow.'

I didn't go to sleep again, but by the time daylight came I'd calmed down a bit. I heard pots rattling — gins putting on the billycans — and then I heard one of them sing out: 'Billycan bin jump up, boss.' I got out of bed and went over. I forget what I did again that day — I probably went for a ride, but by Christ I got a fright that night. If I'd seen a human being I would've shot them.[460]

Although white settlers subsequently assumed the balance of power, the point remains that Aborigines held a considerable advantage for many years. Thus, of necessity, the beleaguered and fearful whites rapidly developed and implemented a range of protective equipment and practices. Foremost of these were always having adequate firearms within close reach, always carrying them out-of-doors and always being prepared to use them. Felix Edgar, the seventy-two-year-old owner of Mount Hart station in the Kimberleys, which Idriess accurately described in the 1930s as a frontier property, unfailingly strapped on his revolver when leaving the homestead, for some of the station Aborigines were 'sulky fellers.'[461]

On a somewhat more bizarre note, Searcy details the usual attire of E O Robinson, 'manager of the Coburg Cattle Company, buffalo-shooter, trepanger and customs officer.' Some years prior to Searcy's visit in 1883, Robinson had returned to his homestead at Croker Island off the coast of the Northern Territory, to find his mate Wingfield had been murdered and buried in his absence. Their fowls had scratched the sand away and exposed Wingfield's face, then commenced to peck it to pieces. Not surprisingly, Robinson was particularly vigilant thereafter and, although he usually walked around naked, he was always dressed with a 'strap and a revolver.'[462] Less than twenty years later Jeannie Gunn, newly arrived at Elsey homestead from her sheltered life in Adelaide, was not only given a revolver for protection but also made to practise firing it.[463]

While such attitudes and actions have been scathingly condemned by revisionists, the most charitable interpretation that can be drawn is that their work lacks honesty. Quite simply, the actions of yesteryear cannot be judged by today's standards, nor from the safety of Sydney or Melbourne more than a century after the event. These settlers had almost no option. A single error of judgement, no matter how well meaning, could see them, their wives and their children massacred. As recently as 1929, for instance, the Walmulla people in central Australia murdered a prospector, made several attempts to kill H Tilmouth of Mt Dennison station, and tried to murder W Morton of Broadmeadow station. On this occasion a party of Walmulla Aborigines was gathering on Morton's property. Being familiar with their language, Morton:

> spoke to a native who had been employed at casual work on the station, and from the demeanour of the

latter suspected that trouble was brewing.

Instead of camping near his soakage or well, as usual, Mr Morton went into open land about 100 yards from the timber. He hobbled his horses close by. Three natives subsequently visited him. One asked for meat. Mr Morton was about to hand some to the aborigine, who unexpectedly gripped him by the right wrist.

On endeavouring to wrench himself free Mr Morton felt himself gripped by the left arm. Being an exceptionally powerful man, Mr Morton felt confident of dealing with three unarmed natives.

When he was seized the other two aborigines, who were about 20 yards away, came quickly on the scene, followed by several armed natives, who had emerged from the scrub. Mr Morton kicked one of the unarmed men, rendering him practically helpless. While handling the other two, who were endeavouring to throw him to the ground, Mr Morton was struck on the head by a boomerang. He then started to fight his way to a blanket beneath which was his revolver.

Anxious moments followed. Mr Morton was several times struck fiercely on the left forearm. With much difficulty he reached the blanket.

His assailants were then striking blows on his right arm and hand, breaking the thumb. Blood was flowing freely from his head. One eye was almost closed. He next received a severe gash on the left side of the chin from a weapon, and another on the right cheek.

As he succeeded in reaching his revolver, Mr Morton was hit on the back of the head with a boomerang. His grim fight for life was not yet over. A burly native was about to deliver a heavy blow on his head, when Mr Morton fired his revolver. The bullet entered the stomach of the native, who collapsed and died.

When Mr Morton fired another aborigine dealt him a violent blow on the head. This brought him to his knees. He again fired the revolver several times, causing his attackers to retreat. They ran into the bush.

In the meantime his horses had stampeded, and after attending to his wounds as well he could he overtook the animals and brought them back to his camp.

Tearing his shirt into ribbons, he bandaged his wounds and rode to the homestead, where his injuries received attention from Mr Sandford. It was subsequently found necessary to reopen a severe head wound from which 15 splinters were removed ...

[Subsequently, police action determined that] the purpose of the Walmullas was to 'kill all whites and station boys along the Lander River, also their cattle.'[464]

Yet carrying firearms did not necessarily guarantee safety, especially in the first half of the nineteenth century, when muzzle-loaders were used. These firearms had severe limitations, the greatest being that they frequently misfired. Misfires arose from a number of causes, including damp powder and percussion caps[465] or, on earlier flintlock models, the failure to generate a satisfactory spark. Furthermore, even for experienced men they were slow to load. When Leichhardt and his party were attacked shortly after dark, near the Mitchell River on the Cape York Peninsula in 1845, no firearms could be fired in response until the weapons had been primed with caps. Once he had found the caps and provided them to Charley and Brown, they 'discharged their guns into the crowd of natives, who instantly fled ...'[466] In the meantime, however, Gilbert had been killed and Roper and Calvert had been speared several times and severely beaten with waddies. Although an immediate volley would not have saved Gilbert, Calvert and Roper would almost certainly have been spared considerable injuries. Indeed, the remaining expedition members were fortunate to have survived the attack. Searcy best captures the intractable nature of this problem in his appraisal of Aboriginal attacks at the early Northern Territory settlement of Port Dundas.

> The question arose how many spears a nigger could throw while a soldier was loading and fixing up his old musket. Those who could have answered were all dead.[467]

Another serious shortcoming of these firearms was that they had only a single-shot capacity. Mention of this problem was made as early as 1800, when Sir Joseph Banks' private collector and natural historian George Caley stressed the virtue of double-barrelled pistols. These weapons, he wrote, are:

> very useful in the woods ... as many of the natives are well aware that a gun will only go off once; nay, many of them do not mind a single gun now ...[468]

These problems aside, smooth-barrelled, muzzle-loading firearms were also inherently inaccurate. Their projectiles lacked the aerial stability imparted to bullets spinning from a rifle barrel and the inefficient powder then available produced a five-foot drop in trajectory over two hundred yards. Moreover, the jolting which occurred in travelling across unmade country meant that the various components of musket powder were differentially distributed in kegs and shooters' horns, resulting in burning rates and therefore trajectories being inconsistent. On top of all this, the sights then available were rudimentary. These factors were offset only by slowly acquired experience and intuition.

Perhaps the clearest example of the inefficiencies of early firearms can be found in George Boxall's *The Story of the Australian Bushrangers*. In the Hunter River district in 1833, a group of escaped convicts ambushed their former master and fired at him repeatedly from a distance of ten yards as he stood in the sheep wash-pool. He escaped unscathed, yet when captured these men were found to possess 'four double-barrelled guns, two single-barrelled fowling pieces, a musket and two pistols.'[469] Overall, the inherent inaccuracy of muskets was such that, even up to the middle of the nineteenth century:

> firing at ranges much over 100 yards was usually a waste of shot and powder. Even at this range the musket was unreliable; its only effective use was to pour volleys into massed troops at very short range.[470]

Clearly, these conditions were rarely applicable in outback Australia, given the Aborigines' previously outlined mode of warfare.

Not surprisingly, cattlemen devised a number of strategies to overcome these shortcomings. North Queensland pioneer Bob Gray of Hughenden, who had served with the British cavalry in India, employed the same aggressive tactics in northern Queensland by sending his:

> mounts as hard as they could with no regard for safety or the slightest hesitation. It worked well with the blacks that used to raid his sheep and cattle. If he caught any of them at it he'd charge like a demon and gallop over them before they had time to think. Their main weapons were spears and these take a few moments to get ready ... Even when ready, it was most unsafe for a warrior to stand there till the charger got within spear-throw of him. If the spear did hit the horse or rider, he'd still get run over. They very soon learned to scatter when they saw him coming and did not like that style of attack at all. They had no counter for it ...
>
> Gray's theory was that it was a mistake to allow blacks to think that white people were helpless without firearms. So many of their actions showed that this was their opinion, but the unorthodox methods of hostility used by Gray set them a problem,

the best solution of which was to give him and his cattle a wide berth.[471]

The effectiveness of firearms increased markedly from the 1860s onwards with the development and general introduction of barrel-rifling, integral cartridges and rotary mechanisms. These improvements allowed considerably longer shots to be made to keep Aborigines away. As early as their surveys of the Darling River region in 1865, Ernest Giles and his assistant W H Tietkens were equipped with modern Schneider rifles, which fired integral cartridges. These .577-inch hollow-nosed cartridges turned 'inside out when [they hit] anything' and the wound bled freely,'[472] and were accurate out to 200 to 250 yards. Furthermore, rapid volleys could be fired at close range with devastating effect. Giles purchased a formidable arsenal for his second expedition and easily repulsed Aboriginal attacks. His lieutenant Tietkens laconically describes one situation, pointing out how:

> The breech loading rifle and revolver had again sent them off with serious loss … [although] several spears were thrown in the dark. I rushed at them and fired point blank emptying the six chambers of my pistol. A stampede followed and I returned to reload … I [later] found a considerable quantity of blood but no blacks.[473]

A cautionary note is required regarding the reliability and efficiency of revolvers, however, for as the early twentieth century cowboy Jo Mora points out, while revolvers were formidable weapons, the caps initially used in the US rapidly fouled and corroded barrels. Furthermore, the open holsters favoured by the cowboys allowed their weapons to become 'rusty and full of dirt and sand.' In conjunction, these factors rendered revolvers 'certainly no precision instrument.'[474]

Revolvers also lacked the range and hitting power of rifles. As the editor of the *Northern Territory Times* pointed out in 1885:

> The chief trouble of the squatter opening up new country seems to be the hostile attitude of the natives, who are described as powerful, numerous and treacherous, so much so that Mr Powers' black trackers will have to be supplied with Snider rifles to cope with them. Mr Davys says one might as well give the niggers homeopathic pilules as fire at them with a revolver. They look upon a revolver as a toy, and it is only at very close quarters that .450 revolver cartridges will do any execution.[475]

Thus it may be that claims regarding the value of revolvers in the outback have been overstated.

The introduction of the Native Police into what was to become southern Queensland by Frederick Walker in 1848, following the considerable success of similar forces in Victoria and New South Wales, was also an extremely important factor in the struggle for safety and supremacy. This paramilitary force, which in 1861 numbered only 120 troopers, was organised into small divisions in which two or three European officers and subalterns each commanded four to six Aboriginal troopers. These self-contained forces were located in various districts, according to demand, and gradually quelled most Queensland Aborigines in a manner entirely disproportionate to their size. Furnished with horses, firearms, supplies and intelligence from local graziers, as well as the black troopers' bush skills, they were formidably efficient. They broke up large assemblies, subjugated the Aborigines through constant patrolling, responded punitively to outrages committed against settlers and their property, and captured suspected criminals.[476]

The officers had absolute authority, to the extent of executing black troopers, and their actions were rarely investigated. As historian Noel Loos comments, whereas the murder of Aborigines in settled districts was a crime, in frontier regions it was regarded as legitimate and defensible. Consequently, when the first duty of breaking up assemblies was subverted to the twin policies of dispersal and keeping-out, little was said. Dispersal, in effect, meant continually breaking up any gatherings of Aborigines, no matter how small, by intimidation or elimination. Keeping-out entailed permanently forcing Aborigines well away from homesteads and regions where livestock were running. This was seen as necessary, for settlers were fearful Aborigines would otherwise recognise their numeric superiority and become emboldened. The general rule was:

> never to allow them near a camp, out-station, head-station, or township; consequently they were hunted by anyone if seen in open country, and driven away or shot down when caught out of the scrub and broken ground. This course adopted by the early settlers and pioneers was unavoidable and quite necessary under the existing circumstances.[477]

Thus stockmen and Native Police harassed and pursued Aborigines continuously, using whips and firearms indiscriminately. Clearly white settlers and their possessions were of paramount importance. Indeed, following the Hornet Bank and Cullen la Ringo massacres in 1857 and 1861, the notion of protecting Aborigines was scarcely considered.[478]

As a result of these interlinked policies, Aborigines were driven from the more fertile areas to densely wooded or broken country, where food was less plentiful. Consequently, the very old and young suffered severely. Furthermore, tribal Aborigines were in large part precluded from moving about their country in order to carry out essential ceremonies. And Native Police troopers sexually abused local Aboriginal women with impunity. Not surprisingly, once this struggle had begun, the frequently starving and intensely embittered Aborigines responded with greater violence, and the relationship worsened inexorably. Sporadic warfare continued in various districts for as long as twenty years, depending upon the nature of the terrain, until the Aborigines were eventually subdued. At this stage pastoralists, who had come to realise the value of Aboriginal stockmen, would allow them to come in and reside permanently at the homestead.

Whether, as Loos suggests, a less combative approach to this clash of cultures would have prevented much of the bloodshed and protracted misery suffered by both sides, will never be known. He mentions two station owners in Queensland who refused to allow the Native Police on their properties and successfully eliminated conflict by providing local Aborigines with a regular supply of food.[479] This approach was not infallible, however, for the Scott brothers at the Valley of Lagoons station on the Burdekin River could not prevent stock spearing by distributing food.[480] Similarly, in 1901 Victoria River Downs in the Northern Territory attempted:

> to check the destructive tactics of the natives by killing a bullock weekly for the sole use of the blacks on the run, but this policy is said to have proved practically useless, and cattle continue to be speared all over the station.[481]

It may be that in the long term, as in the US, a struggle over dispossession was inevitable and that, should the Native Police not have been available, settlers would have carried out the same policy with no restrictions whatsoever. In the wake of the Hornet Bank massacre, for example, local graziers, dissatisfied with the response of the Native Police, formed vigilante groups. At the same time Walker, now disgraced for indiscretions committed when he headed up the Native Police, formed a private security force. Assisted by Aboriginal trackers from local stations, these groups pursued the Jiman people relentlessly and, although within four months the Taroom region was 'virtually cleared of hostile bands' of Aborigines, roving bands of vigilantes continued to traverse the district for another twelve to fourteen months.[482]

Dogs also played a crucial role in securing safety, both around homesteads and further out. A variety of breeds were used, with blue heelers and bull terriers being popular even as early as the late 1800s.[483] On one occasion at Bullita station in the Northern Territory during 1910, the manager, Harry Condon, was asked by an Aboriginal woman to inspect another female who was purportedly ill down at the waterhole, two or three hundred yards away. The ever-suspicious Condon was eventually inveigled into walking down to check her condition. First, however, he buckled on his revolver and whistled his dog. As he got closer:

> more or less opposite one particularly big boab [tree], an Aboriginal with a spear jumped out in front of him. Harry reckoned he got such a hell of a shock he just pulled up in his tracks. The blackfella was fumbling around with his spear and just as he was about to throw it, Harry's bloody dog rushed at him and threw him off-balance.
>
> Harry still got hit — he showed me the scar where he got it right through the bloody arm. I asked him what he did and he said, 'I pulled it out. I didn't know how many blacks were there and I didn't wait to see.' Apparently there was only one, but Harry said, 'I hit me straps for the house as hard as I could go.' And while he ran the cattle dog took after this bloody blackfella and heeled him up while he ran around in circles. Well Harry said it gave him a hell of a fright, but he got through it all right and it was the best thing that could've ever happened, because it made him a lot more careful you see.[484]

Felix Edgar was also completely dependent on his dogs. His homestead was a large, single-roomed dwelling with dried goat- and bullock-skin floor coverings, surrounded by a fence. Idriess wondered how Edgar could sleep at night, knowing that at any time traditional Aborigines or even his station employees could creep in and murder him. The question was answered as he lay dozing on the back verandah, shortly before falling asleep.

> Just then there sounded a low growl, followed by the fluttering pad of a hulking dog as he came across the goatskins inside, then out on to the back verandah. His eyes were glowing. Soundlessly, he prowled on into the yard.
> I understood then. These savage dogs were ceaselessly patrolling: round the yard, in through the always open back door and out through the front, all through the night. No sooner would one lie down than the green eyes of another glowed as he came prowling through the house to the eerie fluttering of the hard, goatskin mats. Those dogs must have saved Felix's life numbers of times.
> I wondered whether, if the natives came and stood motionless, the dogs would recognise them by smell and do nothing. I hardly thought so as another huge brute like a shadowy lion came from the eating-shed. Now and again one would turn on another with throaty growls. From the [Aborigines'] camp still came snatches of wild song and the stamp of feet. No, it is doubtful if even [the outlaw Aborigines] Big Paddy and Possum together would be game to tackle Felix at night.[485]

Savagery towards Aborigines was not an innate trait, though, and these dogs were frequently trained to attack. Reid, the administrator at the Walcott Inlet Government Aboriginal Station in the Kimberleys:

> had trained his [three] dogs to look after Mrs Reid and the youngsters during his absences. Full well those dogs understood their job. Lassie would sleep on the back steps, Jet on the front, Cobber in the yard midway between the steps and the gate. No native dared approach anywhere near after sundown ...[486]

Plate 56: Rainworth homestead, Springsure, built in 1862.

The design and construction of huts and homesteads were also important safety factors. In many instances early dwellings were stoutly constructed buildings with shuttered windows and barred doors. A well-known example is the Rainworth homestead, six miles south of Springsure in Queensland, which was constructed in 1862 following a massacre on the adjoining property, Cullen la Ringo. This homestead is constructed of basalt boulders bound together with a mixture of sand, limestone and crushed calcified basalt, which was dug from the bed of the nearby creek. As safety was a major consideration, the lower windows were fitted with steel bars and heavy shutters (Plate 56). The rear one-third of the building, which was not floored, contained stables and feed and tack rooms, so the horses could be safely locked away each night. The very small windows in the loft are narrow openings approximately six inches wide and may have been intended to serve as rifle slits as much as means of illumination. These factors, in conjunction with a heavy iron roof, meant the building was almost fireproof and largely impregnable.

Camping at night was often extremely dangerous, and on one occasion Searcy and his friends terrified one of their companions by trading on his justifiable fear of an attack by Aborigines. They waited until he had crept into his mosquito net and carefully tucked under all the edges, before loudly shouting a warning.

> The startled one made most frantic efforts to find the exit of his net, but it seemed to have disappeared. He tried all over, top, bottom, and sides, meanwhile performing the most wonderful acrobatic feats. In a short time he so mixed himself up with his net that he could hardly move. If ever there was a properly trussed man, there was one then, and if niggers had been really on the job, poor Footsack would have had a bad time.[487]

Generally, subterfuge was resorted to when camping in areas where Aborigines were known to be present.

> When camping in bad nigger country the usual custom is to sling the [mosquito] net, light the fire, and boil the billy, and then sneak off some distance [in the dark] and camp. Over and over again have fellows found spears through their nets when they returned in the morning.[488]

Homesteads, Yards and Fences

For a variety of reasons, homesteads, yards and fences in the north of Australia have often differed markedly from those further south. By and large homesteads underwent three distinct stages of development. The first buildings, which were usually erected very shortly after arriving at a new property, were generally crude dwellings of one room which provided protection for supplies from the weather and Aboriginal depredations. Often they were constructed of bark or thatch placed over a light framework and secured with greenhide strips. The first dwelling on Campbell's Pilbara property Weelawarrana, which he built in the 1930s, exemplifies this first building phase. It was:

> a twenty feet by ten feet shed [of poles] lined on the walls with spinifex kept together by rails. Later on some old [galvanised] iron from Mundiwindi station would make a roof.[489]

*Plate 57: Delamere station, kitchen and thatched 'dining room',
2 September 1921.*

Where safety was the major consideration, these buildings were generally constructed of solid wooden posts, with heavy shutters and doors. The posts were either set upright in the ground adjacent to one another or stacked horizontally between vertical inner and outer corner posts, log-cabin style. The whole construction was usually twitched together with heavy fencing wire.

With increasing prosperity, available time and perhaps a spouse, more sophisticated dwellings were constructed. Although these buildings employed a variety of design principles in order to make them more comfortable, pragmatism reigned. Homesteads at the turn of the century were usually rectangular and consisted of two or four rooms. One or two were used by the owner or manager and his wife to store clothes and private papers, while the others served as storerooms, sitting rooms or dining rooms. The kitchen and men's dining room were almost always separate, as shown in Plates 57 and 58, to reduce the heat and odours of cooking, and to keep staff away from the house. Sometimes these rooms were attached by a covered way (Plate 59), or located underneath the house. These second-stage homesteads were usually surrounded by wide verandahs, where the owner and his

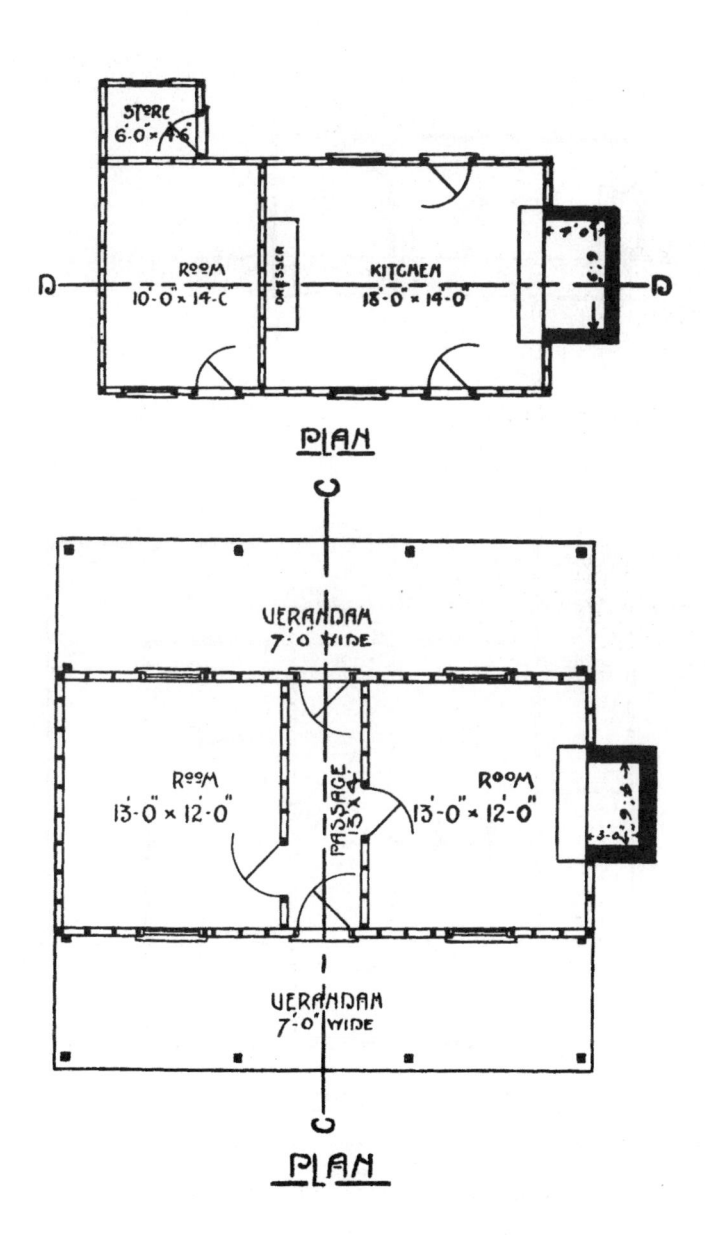

Plate 58: Plan of two-bedroom homestead with detached kitchen.

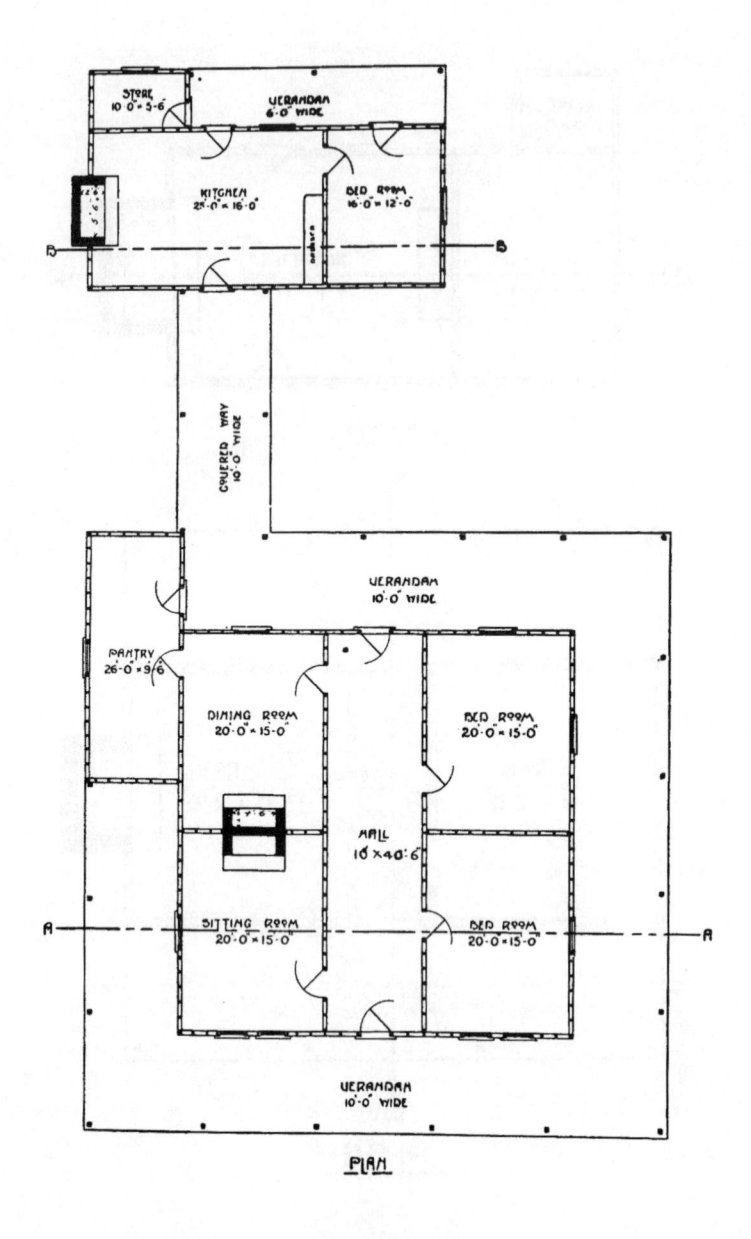

Plate 59: Plan of four-room homestead with detached kitchen linked by covered way.

family slept and ate for much of the year. Partitioning screens were often used for privacy.

The origins of verandahs in Australian housing are debatable. Claims are made that they derive from colonial housing in India, but there is no firm evidence for this. Certainly many military personnel in New South Wales during the early 1800s had served in regions where verandahs were commonplace, including North America, the East and West Indies and the Mediterranean. Whatever their origins, it appears the first verandah in the colony was constructed for Lieutenant-Governor Major Francis Grove in 1793.[490] From Bligh's time onwards verandahs were an established feature.

Setting houses five to eight feet above the ground on stumps also became common in Queensland during this stage of development. There is evidence that this style of building was being erected in Brisbane in the 1860s, where it was appropriate to the steeply sloping terrain of much of the town. Other reputed advantages of this style, which became known as the Queenslander, were that any available breeze would be caught and that air circulating under the house helped keep it cooler. Furthermore, the elevation of the dwelling provided it with protection from flooding in the wet season, extra space for storage, and a degree of security for women and children. Other advantages were some protection from snakes and the existence of a cool space under the house.

Whatever form was used, these buildings often had no ceilings, and the high internal walls stopped well short of the roof. Furthermore, in most cases they had a central breezeway. Ellendale homestead, sixty miles west of Fitzroy Crossing, with doors to each room and a central door on each wall, appears to have had two right-angled breezeways.

Although Ellendale, with its rectangular four-room design and detached kitchen (Plate 60), appears to typify the above, there are some interesting differences. It is constructed entirely of steel and galvanised iron and, because there are no wooden rafters, the roof is held down with hooked bolts. Furthermore,

Plate 60: Ellendale homestead, with detached kitchen.

the floor of the surrounding verandah is a thirty-inch concrete slab. Most surprising of all, the house was imported from Britain in the early 1900s.

Kit houses tend to be thought of as a recent phenomenon, but prefabricated buildings were in fact a feature of Australian housing from the early 1800s, when wooden dwellings were imported from Britain and the US. By the 1850s, however, Australian firms had seized the market and imports almost ceased. An advertisement in *The Stockowners Guide*, published in 1912, shows an example of the small prefabricated wooden homes then available from the manufacturers in Sydney (Plate 61).

Local materials were often used during this stage of development and stone was not uncommon. The original homestead at Strathdarr, on the Darr River north of Longreach, was built before 1890, using local stone cemented together with ground limestone from the riverbed. Similarly, Dorisvale homestead, south of Darwin, was built partly of stone in 1913, using antbed as a matrix. Store and harness rooms were also often constructed from local stone, especially flagstone-type materials.

Campbell also describes how his innovative partner Billy devised a technique for forming mud bricks to build a hut, using an old wooden ammunition case, red Pilbara soil, which has a high clay content, dry wind-grass and water. He soon realised how slow and laborious the technique was and 'hoped the shed would be a small one.' Either his wishes were granted or the process was slower even than Campbell thought, for after

— Simplex Portable Cottage —

——— Perspective View ———

"SIMPLEX" PORTABLE BUILDINGS

are unsurpassed for Shearers' Huts, Boundary Riders' Homes, Station Buildings, Isolation Wards for Hospitals, Schools, Week-end Camps, &c., &c. Our new Catalogue, just issued, contains a large range of Buildings, up-to-date designs, of neat appearance. The cost of erection is very small, and can be carried out by any inexperienced person. Full particulars mailed free on application. We are the Sole Manufacturers of these Buildings for New South Wales.

SAXTON & BINNS Ltd., TIMBER & JOINERY MERCHANTS, PYRMONT, SYDNEY.

Plate 61: Prefabricated home, ex Sydney, 1912.

two weeks' work the shed measured only ten feet by nine.[491] Other examples of pise construction can be found south-west of Winton, towards the channel country, where timber was not readily available. Former Kimberley stockman Joe Moore also recalls pise buildings in Derby and at the Seven Mile at Wyndham, where a Mrs Cole cooked for drovers and teamsters. Overall though, stone, mud brick and antbed brick homes with thatched roofs were not common in northern Australia, for these materials tend to retain their heat throughout the night.

Timber was preferred, for it dissipated heat more quickly and was usually readily available. Another advantage was that timber dwellings could be dismantled and moved. The original Fisher and Lyons Marrakai homestead, which was built around 1876, was shifted three times in an attempt to find a location that was not isolated in the wet season. On each occasion everything was taken but the stumps.

Both round and squared bush timbers were used. On the original Strathdarr homestead all exposed timbers such as door and window frames were adzed square, while the rafters and roof purlins were untreated boree rails. Split slabs were also commonly used for walls, though the method entailed considerable work. Whether vertical or horizontal slabs were used, the principle was the same. Tree trunks were firstly cut to the required length, usually three feet for horizontal slabs and seven to nine feet for vertical slabs, and the round sides split off lengthways with axes and splitting wedges. Substantial corner posts and intermediate uprights were then set two to three feet into the ground, and rammed hard. If horizontal slabs were used, the intermediate uprights were three feet apart. A grooved piece of wood was then attached to each side of the uprights and the ends of the slabs bevelled with an adze so they would slide down into the grooves. Plate 62 illustrates this approach. By stacking the slabs successively on top of one another, the walls were filled. As Jeannie Gunn recounts, the finished product was 'artistic in appearance — outside, a

Plate 62: Cairdbeign homestead, Springsure, showing slab walls.

horizontally fluted surface, formed by the natural curves of the timber, and inside, flat, smooth walls.'[492] Conversely, if vertical slabs were used, the grooved piece formed the bottom and top rails, and the slabs were slid in from the sides.

When using timber, however, the destructiveness of white ants had to be taken into account. These seemingly innocuous grub-like insects have the ability to ravage most materials, including clothes, boots and books. Even living trees are not immune. Before the turn of the century, Springvale station on the Katherine River established an orchard containing 'custard apples, persimmons, mangoes, guavas, lemons, oranges, bananas and pawpaws.' Within a few years only the pawpaws and bananas remained.[493] Perhaps the most graphic demonstration of these pests' voracious appetite is Searcy's account of their eating through thirty-two layers of thin sheet lead in order to get to the soft pine on which it was rolled.[494] Understandably, once established in a building, they can rapidly cause serious structural damage. On several occasions Searcy's chair legs slowly sank through the floor of the Darwin customs house and, 'as can be imagined, strong language was used.'[495] Schultz commented that the best timber available on Humbert River station was bloodwood, but that it did not last long and he was 'sick and tired of putting new Bloodwood rails in my yards every second or third year.'[496]

While the obvious solution was to construct buildings of materials impervious to attack, this was not always possible. Steel framework and galvanised iron cladding were often too expensive, or simply not available. Furthermore, while timbers such as ironwood, Leichhardt pine, palm tree, coolibah, lancewood, woolly butt, cypress pine and boree were resistant to white ant attack, they too were not always available. Accordingly, other strategies were adopted. Firstly, most buildings, especially in the northern regions of the Territory, were built on stilts, 'no matter how high, to keep you away from termites.'[497] The stumps were almost always ironwood and, to improve their efficiency, arsenic powder was placed at the bottom of the stump holes.[498] These measures were highly successful, for the stumps from the original Marrakai homestead can still be seen. Secondly, a galvanised iron cap was placed on top of each stump. These had a one-inch lip bending downwards at forty-five degrees all round, which the white ants could not traverse, thus preventing them from gaining access to the floor bearers.

Though bark and thatched roofs were still in use at the turn of the century, galvanised iron was becoming increasingly common. Not only was it resistant to fire, vermin and white ants, drinking water could be caught from the run-off. Thus, whereas in 1864 only 10 per cent or 183 recorded dwellings in Queensland had iron roofs, by 1921 more than 90 per cent had.[499]

False ceilings were installed using whitewashed hessian, which was either tacked flat to the roof beams or allowed to droop in graceful billows. The seven-roomed Cairdbeign homestead, an expensive building for its time, which has since been transported and re-erected on Rainsworth station, used whitewashed calico instead of hessian.

Floors were often constructed of antbed or cow manure, which set extremely hard, or slate or limestone flagstones. The latter, which are common in creek- and river-beds throughout northern Australia, were sometimes set in an antbed matrix. In short, with high stumps, wide verandahs, high (or non-existent) ceilings, timber-and-iron construction, breezeways and the external walls stopping one to two feet short of the roof, these buildings were exceedingly airy and cool.

Gunn describes this layout at Elsey station. Or rather the proposed layout, for, after the original homestead was demolished by a cyclone, a travelling Chinese carpenter was engaged to rebuild it.

> His plan showed a wide-roofed building ... with two large centre rooms opening into each other, and surrounded by a deep verandah on every side; while two small rooms, a bathroom and an office, were to nestle each under one of the eastern corners of this deep twelve-foot wide verandah.[500]

Unfortunately, while the plan was eminently suitable for tropical life, the carpenter failed to calculate the quantity of timber at hand, and consequently ran short well before completion. He used 'joists and uprights with such reckless abandon that by the time the skeleton of the building was up, the completion of the contract was impossible.' Thus, as Mac the stockman put it, the house was 'mostly verandah and promise.'[501]

In the third stage relatively sophisticated homesteads were built, although the basic design principles remained the same.

While stone buildings were constructed, such as those at Coolibah, Glenroy, Liveringa, Mount House and Humbert River stations, they were not common. Iron roofs and sawn timber were preferred, although prior to the late 1800s that often involved pit-sawing. Once steam-powered sawmills were established in Brisbane in the 1860s and the rail system expanded throughout Queensland, sawn timber became much more readily available. Such was not the case in the Kimberleys and the Territory, where few lines existed. Certainly, all materials for the Inverway homestead, east of Halls Creek, were shipped from Perth to Wyndham by the State Shipping Service in about 1917, then carried hundreds of miles by pack camels.

In fact transporting commercially prepared construction materials was often a considerable problem. In 1906, when Frank Dean's grandfather built his homestead on Notus Downs, south of Longreach, the materials were carted by bullock wagon from the railhead to within ten miles of the building site. Frank's grandfather then had to carry them the remaining ten miles. As the longest pieces of sawn timber were twenty-six feet long and his dray only eight feet long, preventing the timber protruding forward and interfering with the horses seemed an insoluble problem. With bush ingenuity, he loaded the galvanised iron on the floor of the dray, then the shorter lengths of timber and finally the longest pieces on top. With this elevation they were well clear of the horses' backs.

Gradually more and more outback stations acquired bench saws and cut their own timber. Annette Henwood of Fossil Downs recalls vividly her father Reg McDonald cutting the timber for their new homestead in the early 1940s, and how the Leichhardt pine was bright orange in colour. Initially these saws were powered by steam and later by petrol engines. Schultz overcame his lack of an engine in the late 1940s by jacking up one wheel of his lend-lease truck and using it to belt-drive his sawbench. Sawmilling brought new problems, however, for huge logs had to be transported to the sawbench. According to Henwood, after fifty years 'there are still spots coming in from

Plate 63: A Nissen-hut-style homestead at Roebourne.

the river where you can see the wheel ruts where the donkey wagons brought the timber in.'[502]

Although the above scheme of development is generally applicable, some regional variations of design occurred. Foremost of these was the Nissen-hut style of construction, which appeared in Western Australian homesteads and sheds from the early 1900s. The advantages of this form of building were cheapness and simplicity, for no roofing framework was necessary. Pre-curved sections of galvanised iron were packed in tight bundles, shipped northward by the State Shipping Service and carted inland. On arrival, the sheets of iron were bolted together and erected. Unfortunately these buildings had one inherent fault: they were exceedingly hot, being inherently low structures due to the curvature of the iron. More sophisticated versions had the curved iron mounted on top of wall beams, with verandahs surrounding the core building. Plate 63 illustrates this, as well as the detached kitchen and showers.

Variations in design were also influenced by a cattleman's commitment to the industry. Reg McDonald of Fossil Downs saw his life's work in the Kimberleys and decided to build a substantial and attractive dwelling. He commenced construction on New Year's Day 1939, and building continued intermittently for the next eleven and a half years, being interrupted by World War II, wet seasons and everyday cattle work. Firstly, he shifted the site, for not only had the original homestead been skewed sidewards on its stumps during the 1914 flood, it was on the edge of a large blacksoil plain, and blacksoil becomes intolerably glutinous after even a light shower of rain and sticks to boots and clothing. Furthermore, at night 'every wog for thirty miles could see [the lights].' The present site was chosen on the advice of local Aborigines, who declared it was beyond the reach of floodwaters. This proved to be the case in McDonald's lifetime, but in the past fourteen years it has been flooded five times.

McDonald, who designed and oversaw the construction of the two-storeyed homestead, chose to build in cement brick. He calculated that ten thousand bricks would be required and established a mixing site under a bough shed in the riverbed, where washed sand and water were readily available. As the bricks were removed from the moulds they were placed in water to cure, then stacked on the riverbed. According to Henwood:

> the day they had them all made he and my mother decided to go into Fitzroy, and then on down to Nookanbah, to my husband's parents' property, to celebrate completing the bricks. They got as far as Fitzroy Crossing and the publican stopped them to tell them the MMA pilots had flown over Fossil and could see no-one was there, so had dropped a note at the Crossing Inn on the off-chance dad was in town, to say the river was coming down. They turned

around and came racing back. All the natives living here and my parents worked for 24 hours. My mother said their hands were raw and everybody was practically in tears from exhaustion, but they got the 10,000 bricks out ... My father [then] decided to go upstream to see where the river was. They only went a few miles upstream and saw it coming, and raced back down to the bough shed. Mother said it was the only time in her 50 years up there that she saw the river come down in a big wall. The huge cement slab that was underneath the bough shed went end over end and they never saw that again.[503]

In these two later stages of development a variety of supplementary devices, including insulation, were used to keep houses cool. Coarse local cane grass was used on the original Strathdarr homestead. Holes were drilled through the rafters eight to nine inches apart and wire run through. Cane was then laid five to six inches deep on the wire and the galvanised iron screwed down over the top with heavy screws. According to Dean, the building was 'beautifully cool.' At pioneer Jim Fleming's homestead on the Douglas River, south of Darwin, stringy-bark was stripped, heated and flattened, then nailed to the bottom of the ceiling beams with the smooth inside facing downwards. This not only provided a ceiling, but made the house quite cool. Similarly, Gunn outlines how 'great sheets of bark, stripped by the blacks from the Ti-tree forest, were packed a foot deep above the rafters to break the heat reflected from the iron roof ...'[504] *The Stockman's Guide* mentions that using a layer of charcoal in the roof effected wonders in keeping houses cool. Thatch was often used to keep meat houses cool (Plate 64).

Split-cane lattices, or fixed or movable wooden shutters, were also used to enclose verandahs and keep them shaded. In the Pilbara region of Western Australia, where severe cyclones are common, steel shutters were used. These were left propped

Plate 64: Fly proof thatch-roofed meat house.

open until a cyclone was imminent Further north, in the humid tropics, it was not uncommon for large sections of the steel-clad walls to be hinged just under the roof-line and propped open as required. Unfortunately, as the McDonalds found at Fossil Downs, these ingenious and complementary methods could not prevent their brick homestead from gradually absorbing and retaining heat.

> There were no air-conditioners or fans then and these brick houses, once they get hot, they stay hot. So we were pretty grateful for our bough shed.[505]

Bough sheds were an ingenious invention based on the Coolgardie safe evaporative principle. A light framework of timber was covered with wire netting and thatched with spinifex or other native grasses The guttering that ran around the building was punctured at regular intervals so that, as water was piped up into the gutter, it trickled down through the thatching and produced a cooling effect by evaporation. Fossil Downs' bough shed was slightly more sophisticated than most, having a flagstone floor with a drain, and corrugated iron under the roof thatching in order to make it waterproof. In the hot part of the year Henwood, her mother and sister moved into it after lunch each day for two to three hours. When they commenced their schooling they did all their correspondence lessons there and, as Henwood recalls, 'I don't know what we would have done without that bough shed.'[506]

Stockyards also underwent considerable development and refinement. Initially they were constructed of wood and consisted of one or two small, roughly built yards with a wing at the gateway. Probably the most unsophisticated would be that described in 1861 by Ernest Henry of Mount McConnell station on the Burdekin River. Henry, who had arrived from Britain only a few years previously with no experience of Australian conditions, wrote to his mother outlining how he was building a fork-and-rail cattle yard. He firstly cut forked trees to length and set the butt two feet in the ground, then stacked rails horizontally within the forks. Not surprisingly, he mentions that his yard was very roughly built.

Stub-yards, which are made of posts set vertically some two to three feet into the ground alongside one another, were common first yards. Although requiring no skills other than digging ability, they were wasteful of posts and were replaced by post-and-rail yards. These consist of substantial posts, nine to ten feet long, set two to three feet into the ground, with several intermediate horizontal rails for each panel. For many years rails were affixed by wire, using a number of twitching techniques. Plate 65 shows the remnants of crude yards at St Vidgeons station on the Roper River, while Plate 66 illustrates a more sophisticated system, whereby a check was cut from the post and the rail inserted flush, then twitched (wired) around the back of the post.

Over time wire twitching was replaced by bolting, whereby the ends of two abutting rails were cut at an angle and overlapped on one another against the post, then a bolt passed through both rails and the post. Mortising is a more sophisticated method, in which indentations are cut into the post and the ends of the rails adzed to fit. The rails are then inserted into the mortises and, as each panel is assembled, the rails are locked in position (Plate 67).

Plate 65: Remnants of paperbark yards on Saint Vidgeons station.

As properties became more heavily stocked, additional yards were constructed near outcamps in order to hold cattle as they were mustered. Wire yards were often built, for timber yards required considerable labour and suitable materials were not always readily available. Wire yards had timber uprights, but used heavy wire instead of rails. For greater security, wooden cap and belly rails were frequently incorporated. Although easier to construct than rail yards, wire yards were still no sinecure. Each wire hole had to be drilled with a hand auger and, to ease the workload, double-handed, heavy-duty augers were developed. Wire yards had their limitations, however, for wild or badly excited cattle could go through them. As Wilson recalls:

> Those Marakai cattle ... used to jump and were very wild ... [They] would have gone straight through the yards ... if they were wire. Wire was for quieter cattle.[507]

While the wire used was often number eight gauge, twisted into a strand to give greater strength, Wilson recalls seeing the extremely heavy Overland Telegraph wire used on a number of Jim Fleming's yards. This wire, which was about a number two gauge and was difficult to bend and join, was known locally as 'number two bull-wire'. Whether this was intended ironically, or to prevent the authorities hearing of its illicit use, is not known.

Plate 66: Post-and-rail yards showing wire twitching.

Gate hinges also demonstrated ingenuity, especially where a blacksmith was not available. Initially, hinges were of greenhide. With time, the lower hinge was constructed of a white-ant-resistant block of wood set eighteen to twenty-four inches into the ground. An indentation the width of the gate upright was made six inches into the block and the upright set in it. This block then carried the weight as the upright pivoted on it. An indentation in the cap rail, which fitted over the top of the gate upright, formed the top hinge. Alternatively, a four- or five-inch loop was forged on one end of a piece of metal and knocked down over the top of the gate upright. The straight end was then fitted through a hole drilled in the gate post and held in position by a key, or threaded and tightened with a nut.

By the late nineteenth century, yards had been refined considerably, and incorporated scruffing pens, dips, races, and bronco yards and panels. *The Stockowner's Guide*, published in 1912, includes plans for different-sized mobs of cattle. These show mangates placed around the yards wherever they were 'handy for men in getting from one yard to another, and for getting out of danger' (Plate 68).[508] The plans for a set of horse-breaking yards shown in Plate 69 include drafting gates, receiving and forcing yards, and a round yard or 'pound' as it was commonly called. The size of the round yard was often a sore point with horsebreakers: if too large it allowed the horse to buck harder, while too small a yard could frighten it and encourage it to try to smash through the rails. As horses can clamber over a higher fence than cattle, *The Stockowner's Guide* recommended round yard-rails be a foot higher than for ordinary cattle yards. A 'crush' opening into the round yard allowed horsebreakers to gear up or saddle a horse in greater safety.

Plate 67: Remnants of post-and-rail yards showing mortising.

Steel piping began to replace wood during the 1950s, steel being immune to white ants, less liable to break and easier to erect. Perhaps the most whimsical example of Territory ingenuity is the usage of materials from old railway lines for

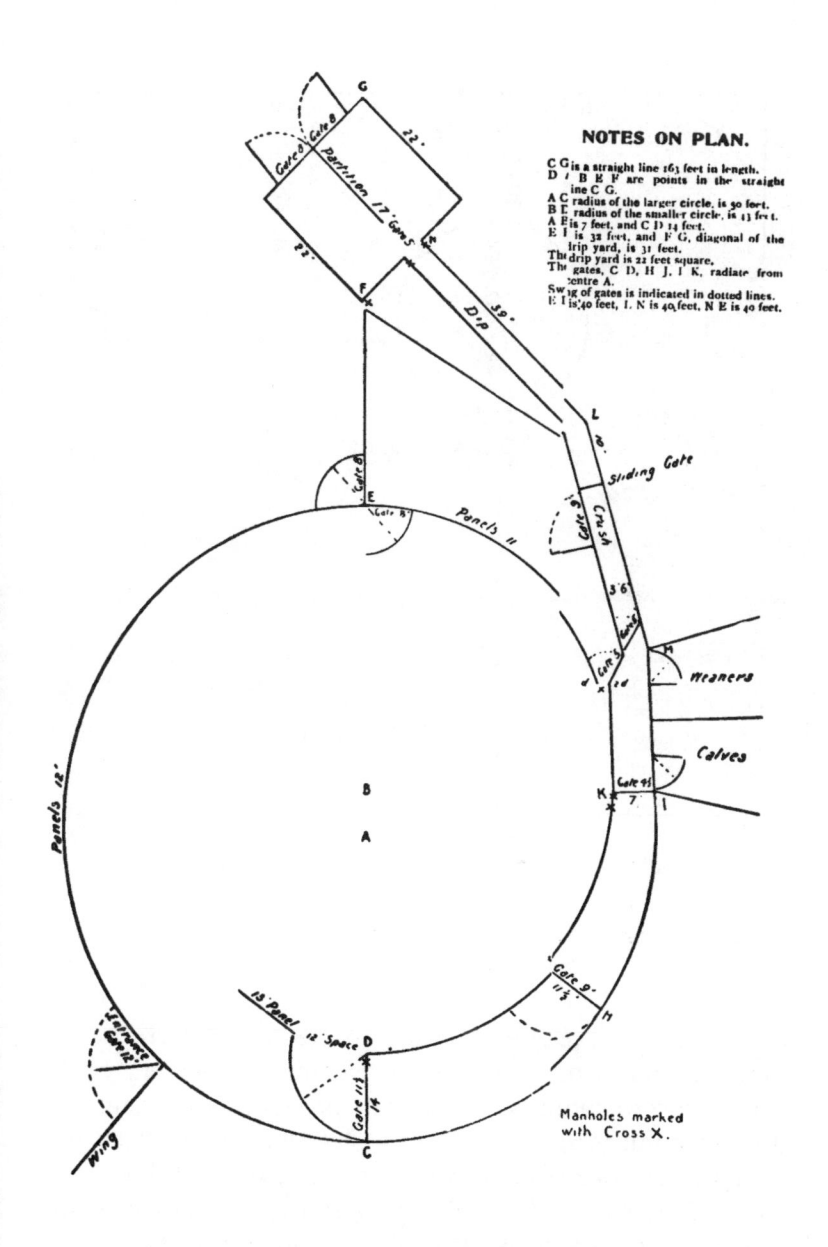

Plate 68: Plan of cattle yards with mangates (marked X).

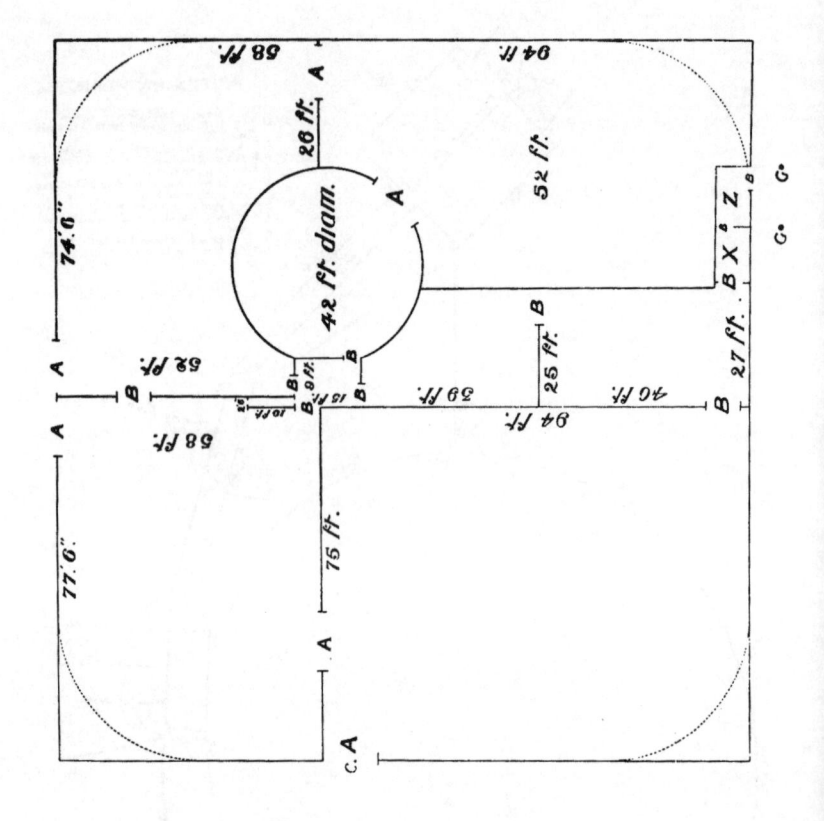

Plate 69: Plan of horsebreaking yards.

cattle yards. For obvious reasons sleepers in the Territory were made of steel, and cattlemen seized their opportunity when lines were decommissioned and torn up, using lengths of discarded rail as uprights and steel sleepers as rails (Plate 70).

Fences in northern Australia initially resembled those in the south. They were very solid, almost rigid structures, consisting of heavy wooden posts set twelve to fourteen feet apart with combinations of several plain and barbed wires. The high cost of transport, the very large distances involved and the fact that only cattle were being run soon led to changes. Cattlemen found that spreading posts much further apart, even up to a chain, reducing the number of wires to three and using intermediate droppers still made for an effective fence, especially if barbed wire was used.

Barbed wire was invented in 1874 by an Illinios farmer, Joseph F Gliddon, in response to the cost of fencing the numerous small properties being developed on the Great Plains. In 1871 the US Department of Agriculture estimated that two billion dollars was invested in fencing nationwide and that the annual upkeep cost two million. Gliddon commenced production in November 1874, hiring a crew of boys to string the barbs on the twisted strands. Business thrived and in 1876 he sold a half interest to the leading wire manufacturer in the United States. This company had already developed a machine

Plate 70: Steel rail-and-sleeper cattle yards at Nathan River station, Limmen Bight River.

to produce the wire by a fully automated process and, within a further twelve months, had fifteen similar machines in operation. By 1882 the Gliddon factory was making six miles of barbed wire from fourteen miles of plain wire every ten hours.[509]

Although highly effective in controlling cattle, barbed wire can be difficult to run out, particularly through bored holes. Cattlemen overcame this by using a bigger auger bit and dragging the wire through the posts with a quiet horse. As the wire was pulled up to a post, the horse would stop while the wire was unhitched, passed through the bored hole and re-attached. It would then walk steadily on to the next post.

Wooden posts were always vulnerable to bushfires, and considerable care and effort were involved in safeguarding them. Frank Dean's grandfather, who was head stockman on Bowen Downs in the 1870s, sent men to check whenever control burns moved across fencelines. Small smouldering sections were cut out of a post with a butcher's knife and larger burning pieces chopped out with a tomahawk. Not surprisingly, fire- and white-ant-resistant steel posts were introduced before the turn of the century. Indeed an advertisement for 'Bain's steel standards' (posts) in *The Stockowner's Guide* shows that Claude Healy and Co. of Sydney began selling steel posts no later than the mid 1880s. They were not widely adopted for some time, however. Dean recalls an early T-shaped steel post being used in the Longreach district of Queensland around 1930, but steel posts were not common in the Northern Territory or Kimberleys until the 1960s. Presumably this had as much to do with the high cost of transport as with the paucity of fencing in those regions.

Rangeland Management

At the turn of the century rangeland management is a contentious topic. Certainly there is no question that grazing has deleteriously affected large areas of the Australian rangeland. On Francis Gregory's 1858 exploration between the Murchison and Gascoyne Rivers, for example, he recorded crossing:

> a succession of stony ridges thinly grassed and nearly destitute of trees; in the valley the kangaroo grass was tolerably plentiful and quite green.[510]

Yet that country today is largely bereft of grass, save isolated areas sparsely covered with soft spinifex and poor quality wind grass. Overall, a 1992 study by Tothill and Gillies found that 32 per cent of the northern Australian rangelands used predominantly for grazing cattle can be classified as degraded but recoverable, while a further 12 per cent are degraded to the point where they are economically irrecoverable. That is, 44 per cent of Australian rangelands are degraded to some extent.[511] Thus it is hardly surprising that state departments of agriculture have legislative powers to compel property owners to partially or even completely destock stations, to allow them to recover from sustained overgrazing.

More importantly, a national approach to rangeland management was mooted at a 1992 meeting of arid land administrators. The resulting Federal Government Environment Policy Statement of 1993 acknowledged that the rangelands required urgent action and the desirability of input from all interested parties. Accordingly, a Rangelands Issues Paper, covering such questions as ecologically sustainable development, international obligations and scientific monitoring, was issued in 1994. The findings of 182 submissions were then collated and incorporated with the material gathered from thirty meetings held throughout Australia. These findings formed the basis of the 1996 Draft National Strategy for Rangeland Management. Eventually, a comprehensive twenty-five-year strategic plan, which embraces as well as possible the often conflicting desires of several disparate single-issue groups, will be finalised and implemented. All land-users will then be provided with:

> certainty of use and access in order to make long term investment and management decisions. In return, they [will] need to be accountable for what they do [to the rangelands].[512]

Concern over rangeland management is not a recent phenomenon, however. There is evidence that as early as the 1860s cattlemen in the Longreach region were attempting to preserve their better grasslands and improve their poorer areas. Unfortunately, while cattlemen generally understood the ideal means of managing their properties, they were rarely able to achieve their goals. In order to maintain production and regeneration, control over watering, grazing and fires is necessary. To some extent these factors are interrelated.

Fire has traditionally been used to clean out old dry grass and stimulate young fresh growth, especially against the very rank grasses of the Top End and the spinifex of the more arid regions.[513] Burning was carried out at the end of the dry season, during dry spells in the wet season, or shortly after the end of the wet. According to Wilson, burning commenced on Burnside and Marrakai during the 1930s:

> straight after Easter and no later than May ... when ground moisture was still sufficient and the temperatures were still high. You got good regrowth [then].[514]

Care had to be taken, however, that the country had not become too dry. Burning on Fossil Downs, for instance, was done 'when it was dry enough to burn but not too dry to literally take off.'[515]

Burning has not been unanimously accepted, though, with many cattlemen holding quite discrepant opinions on the merits of the practice. J H Kelly, for instance, mentions in his 1959 appraisal of the Leichhardt–Gilbert region of north Queensland that one local cattleman:

> with long experience as a station manager and as a supervisor of the management of a group of stations which carried up to quarter of a million head of cattle and branded upwards of 50,000 calves annually, expressed the opinion that the indiscriminate burning of grass ... had resulted in substantially reduced cattle numbers, brandings, and turn-off in the Gulf Country.[516]

Conversely, he reports that another:

> grazier, who is a descendant of the pioneer of one of the earliest cattle holdings in the north, established about a century ago, said that station records show that the carrying capacity was no greater originally than now, despite the fact that the grass had been burned annually throughout the long period of occupancy and cattle grazing ...[517]

Certainly, burning has not been universally practised. Cattlemen on blackplain soils, for example, found that burning destroys the dry plant material which normally replenishes these cracking, self-mulching soils. Furthermore, as stocking rates have increased through the subdivision of Queensland stations and improved management practices, cattlemen simply cannot afford to waste grass by burning it. Another factor is that burning deleteriously affects the better perennial grasses. As Henwood points out, albeit with some qualification:

> Mitchell and Flinders [grasses] ... take a long time to recover. Having said that, we're wondering if we are being over protective of them. In recent times we've had some heavy wets and we've now got a big body of old, dry feed that nothing is eating. We are starting to wonder if it does need a burn, but we are a little bit cautious, because we've always felt these grasses don't take kindly to it.[518]

Accordingly, only discrete areas are burnt, depending on the prevalent local species. On Fossil Downs, one-third of the country:

> is river-frontage and blacksoil plains, which is our best country. This grows Mitchell and Flinders grass on the blacksoil and Buffel, Birdwood and native grasses on the river-frontage. [This isn't burnt.] Then we have one-third which is more Pindan country,

Plate 71: Early station fireplough.

where we are introducing new grasses such as Veruna, Seca and Euroda. It too has good areas of spinifex country [which is burnt selectively]. And the one-third of top-end country, which is mostly rocky and rough country ... that's where we have a lot of spinifex ... We benefit a great deal by burning the spinifex. Cattle really love it when it's kept short and green. When it has had a good burn the cattle do so brilliantly on it.[519]

Indeed, spinifex was often burnt as soon as there was sufficient material to carry a fire. In the Longreach and Cloncurry regions, for example, it was burnt on about a four-year cycle. Having a cycle of this length prevented the soil being cut up by stock and also safeguarded against being left with little feed should the following wet season prove disappointing.

Plate 72: This Wave Hill fireplough required a twenty-horse team (9 September 1921).

In order to control fires and protect the better grasslands, firebreaks were made using primitive graders known in Queensland and the Northern Territory as fireploughs. At first they were made by station blacksmiths (Plate 71), but commercial ploughs later became available. These were constructed of two eighteen-foot planks ten inches high by four inches thick, bolted together in the form of a vee. To the bottom edge of each plank was fixed a serrated steel cutting plate, and the apex of the vee was covered with a steel nose-cutter cap. More sophisticated versions were suspended from a wheeled frame and could be adjusted for depth by a threaded screw (Plate 72). Before the advent of tractors these ploughs, which cut a ten-foot-wide track, were pulled from the front of the vee by twenty-eight horses harnessed four abreast.

By burning three- to four-chain strips back on to these firebreaks during the evening in the cooler months, substantial firebreaks were made. Pat Underwood frequently used this technique on Inverway to protect his better quality perennial grasses from fires that started in breakaway spinifex country after lightning strike. Burning was also carried out on a patchwork basis, so that a fire could not travel too far before it came up against a section that had been burnt in the previous year or two, leaving it with insufficient fuel to carry a fire.

Ironically, as fire control techniques have improved, many cattlemen have found their properties invaded by what are known as woody weeds. These are opportunistic trees or tall shrubs which have spread far beyond their previous boundaries. Not only do they shade out grasses, thus reducing carrying capacity, they can also make mustering extremely difficult. Gidyea (*Acacia cambagei*), which grows to a height of thirty feet, has established dense stands on cracking soils in the Longreach region. A considerable amount of gidyea seed germinated in the 1950s following a series of good seasons and, because the seedlings were not killed by grassfires as would normally have occurred, stands of up to several thousand acres are now commonplace. The situation is similar in the Top End, where Wilson recollects having seen:

> a photo taken by a surveyor's wife, Mrs Briggs, at the Katherine Telegraph Station in 1914 or 15, and it showed virtually no trees. My uncle was in the same survey office and said Aborigines' burning practices at that time had skinned all the smaller trees out and that's why it looked more open. When I first went down there in 1950, working on surveys, it was fairly well-timbered, except for the farms.[520]

Overall, by the late 1980s 17.3 per cent of Australian rangelands were affected by woody weeds.[521]

The capacity to supply water where and when required is also vitally important in managing rangelands. As Kelly points out:

> Adequate water supply is defined as having sufficient permanent cattle watering points to enable the grazing resources to be utilised fully and efficiently. This can be achieved by having watering points so spaced that cattle can graze within three to four miles of water ... to ensure that the pasture around water points is not overgrazed ...[522]

Yet for much of the first century pastoralists had little or no control over water supplies. Very few permanent waterholes existed and droughts were commonplace in inland Australia. Nineteenth century commentator Nehemiah Bartley, for instance, outlines how:

> Captain Flinders, cruising about the coasts of Australia at the commencement of the present century, found, everywhere, the bush on fire, grass burnt and withered, and every sign of great and long-continued drought. In 1828 and 1829 came the drought of the century, with water at fourpence a gallon in Sydney — the great Murrumbidgee River dried up, the fish dead in the dry mud of it ... Then in 1849 and 1850 came the terrible drought, which culminated in 'Black Thursday', in February, 1851, when burnt leaves were blown across Bass Straits by the fury of that north wind, which amalgamated into one huge blaze the previously scattered bush fires of Port Phillip ...
> The question for us now to consider is this: will the great periodic droughts, extending over eighteen months or two years at a time, which has already happened three times in a century, and at apparent intervals of twenty-five years, more or less; will it come again to us, and how soon? I don't think it is on us yet, but I think it is only a year or two away from us — that is to say, if past experience be any guide, and if dependence can be placed on statistics. It is a serious

> matter to contemplate ... The sheep and cattle of 1875
> will outnumber those of 1825 and 1850 by an amount
> so vast as to render the prospect all the more terrible.[523]

In writing the above, Bartley drew from his 1864 paper to the Philosophical Society of Queensland, in which he concluded that droughts and abnormally wet years occurred over twenty-year cycles. In fact the situation was worse than he believed for, in the hundred years from 1860, Australia suffered twenty-seven years of drought, or worse than one year in four. Paterson best captures this constant battle for survival in 'With the Cattle'.

> The drought is down on field and flock,
>   The river bed is dry;
> And we shift the starving stock
>   Before the cattle die.
> We muster up with heavy hearts
>   At the breaking of the day,
> And turn our head to foreign parts,
>   To take the stock away.
>   And it's hunt 'em up and dog 'em,
>   And it's get the whip and flog 'em,
> For it's weary work is droving when they're dying every day;
>   By stock routes bare and eaten,
>   On dusty roads and beaten,
> With half a chance to save their lives we take the stock away.
> We cannot use the whip for shame
>   On beasts that crawl along;
> We have to drop the weak and lame,
>   And try to save the strong;
> The wrath of God is on the track,
>   The drought fiend holds his sway,
> With blows and cries and stockwhip crack
>   We take the stock away.
>   As they fall we leave them lying,
>   With the crows to watch them dying,

Grim sextons of the Overland that fasten on their prey;
  By the fiery dust storm drifting,
  And the mocking mirage shifting,
In heat and drought and hopeless pain we take the stock
    away.[524]

Unfortunately, the problem was not entirely one of drought, for water supplies were often inadequate in years of average rainfall. Palmer writes of:

> the anxiety of the stock owners towards the end of the season, when all surface water (except the most permanent lagoons) has dried up and formed mud traps to catch all the weak stock that venture near them.[525]

In an attempt to overcome this highly dangerous reliance on good rains, Queensland cattlemen initially constructed overshoots across creek channels and suitable riverbeds. Overshoots are walls built downstream from the more reliable waterholes, using the naturally occurring flagstone so prevalent in northern and central Queensland. They bank the water back for a considerable distance and raise its level while still allowing it to overshoot during floods. Often built by Chinese labourers, overshoots were so well designed and constructed that many are still standing in northern Queensland.

In areas where suitable overshoot sites could not be found, excavated tanks (dams) were dug. Firstly, the ground was ripped by heavy mouldboard ploughs drawn by nine horses or bullocks harnessed in four pairs plus a leader.[526] The loosened earth was then collected in scoops, carted to where the wall was to be built and dumped. The earliest scoops, which were drawn by two horses and carried three to five cubic feet, simply slid along the ground and picked up the soil. Later, more sophisticated, wheeled Gaston and monkeytail scoops (the latter so named for their long handles) were introduced.

Usually drawn by five horses or bullocks harnessed abreast in a fan configuration, these carried nine cubic feet of earth.

When no water was available at the work site, teams were camped at a nearby waterhole and walked to work each morning while a temporary hole was dug. It was only after rain had fallen and filled this hole that excavation commenced on the main dam, the temporary hole being used as a source of water. Alternatively, water was carted for the working stock. In 1911, for instance, the bullock team working on an earth tank at Notus Downs was watered by a three-horse team pulling a dray loaded with a four-hundred-gallon ship's tank. Filled by bucketing from a waterhole, it was carted eight miles each round trip.

Unfortunately, these measures were of limited value in bad seasons, and in the 1880s the Queensland and New South Wales governments appointed geologists and engineers to construct large dams and tanks along stock routes so stock could be shifted during dry years. While at first glance it would seem that these improvements, in conjunction with the development of railway lines throughout Queensland, would have contributed to rangeland management by allowing graziers to move stock from drought-affected areas, Frank Dean argues otherwise.

> If anything, it has been detrimental. On our original property, 60 miles south of Longreach along the Thompson River, you had to get stock off the property while there was still water in the main waterholes along the river. Even if you had a bit of pasture left on your place. By the time the stock returned your pasture had recovered to a large extent. Today, they can flog the country to the last gasp, and the day after it rains they can be back again.
>
> If we'd no rain at the end of the wet season in April, you got out. We walked our stock out in March 1938 and couldn't return until April 1939, even though we'd had good rain in November. There was no rain

or feed in between and we had to wait until it rained on the country in between.[527]

Wells were also sunk, but with limited success. Water was found at Winton at 336 feet, for example, but it had to be pumped to the surface and the supply was inadequate. The situation worsened until, in January 1884:

> Winton was only saved from a water famine by trains of tanks mounted on wagons being hauled there by the traction engines used in conjunction with steam scoops for excavating reservoirs.[528]

As a result, graziers and engineers began turning their attention to the numerous permanent creeks and rivers of Queensland, and the smaller number in the Territory, as well as the mound springs at the southern and western shores of Lake Eyre. In 1879 a hydrologist named Tate posited that these springs were artesian in nature, for they flowed to the surface of an open level plain in an arid area, and accordingly he surmised that artesian water could be found in much of central Australia. At about the same time it was realised that the beautiful, tree-lined, permanently running Gregory River, which was 'estimated [to] flow ... at 133 millions of gallons per day at Gregory Downs,'[529] must also be fed by an immense artesian source. Accordingly, interest grew in the French experiments in tapping artesian water.

Curiously, the origins and principles of artesian water were initially poorly understood by the Australian public. Palmer, for instance, despite being in possession of:

> the report of the Hydraulic Engineer, [which was] a coloured map showing the sites of artesian bores and tanks and the supposed areas of ... water bearing strata ... [of] the whole of western Queensland ...

could only comment that:

> the source of this enormous pressure of water that is capable of sending a jet over a hundred feet above the surface, is still unexplained, and many theories are afloat as to its origin; some of these go far afield for reasons for the great supply and strong pressure. The enormous rainfall on the coast ranges, where the intake probably occurs ... seems to be the most reasonable [theory] to adopt at the present time.[530]

Professional geologists such as Robert Logan Jack, however, had no doubt. Records indicate that the first artesian bore in Australia was sunk at Kallara station, near Bourke, in 1879 and Jack suggested as early as 1881 that a large synclinal trough existed in western Queensland, that water from the Great Dividing Range probably lay in the pervious beds of the trough, and that conditions appeared favourable for artesian wells. Within a few years severe drought led the Queensland Government to direct Jack to investigate and report on the potential for artesian water. He quickly undertook a drilling program and concluded that nearly all the western interior appeared promising. Drilling commenced almost immediately at Blackall, which was in a parlous situation, but due to drilling difficulties the 1663-foot bore was not completed until 1888. Meanwhile other deep wells were successfully sunk at Cunnamulla (1290 feet) and Barcaldine (691 feet).[531]

While these dates are documented on Government records, some disagreement exists over the date of the first Queensland bore. Grazier Simon Fraser notes that in 1883 the property Thurulgoona, which he and others owned in partnership, was suffering severely from drought. At this time Fraser met a Canadian driller, J S Loughead, who had come to Australia with his drilling plant hoping to find payable petroleum deposits. Receiving no governmental assistance, he was searching for private work and, on 9 June 1886, signed a contract to bore for water 'in such place or places as [Fraser's] Company ... may direct, to an aggregate depth of 5000 feet, in five or more wells.'

After a delay of some months, while more equipment was shipped from the US and the entire outfit was carted to Thurulgoona via Bourke, drilling commenced:

> on December 17th, 1886 ... in the Cotton-brush Paddock about 30 miles from the township of Cunnamulla, Queensland. At this time the Hydraulic Engineers Department was prosecuting boring operations by means of Tiffin machines, Water Augers, and other appliances, progress with which was slow and unsatisfactory; but Lougheads undertaking was carried out by means of a Canadian Pole Tool plant, the first of its kind introduced to Australia.
>
> In six weeks he had reached the depth of 1000 feet, and had not gone much further when he found water rising in the bore and overflowing from the tubing at the surface of the ground. This event happened in February, 1887, and was the beginning of the artesian water supply of Australia.
>
> ... There was no mistaking ... the fact that artesian water had been struck, but to my dismay, on measuring the supply I found that it only amounted to about 3000 gallons a day — a mere trickle for stock-watering purposes.
>
> Subsequent events proved that the actual supply available had hardly been broached, for the sinking of the same bore another 100 feet increased the flow to 80,000 gallons per day; while the second bore, which at 1241 feet gave 1000 gallons, on being deepened to 1682 feet, was made to yield 576,000 gallons in twenty-four hours.

According to Fraser, Loughead then negotiated with the Queensland Government to drill an aggregate of 7500 feet in three bores of not less than 2500 feet each, and commenced the previously mentioned Barcaldine bore on 16 November 1887. Within a month he obtained a flow of 175,000 gallons per day at a depth of 645 feet.[532]

Whatever the truth of the matter, graziers quickly realised the immense advantages of permanent, potable supplies of water. Not only were the dangers of drought greatly reduced, but carrying capacity was markedly increased. Furthermore, bore-sinking required drillers, steel bore casing, steel cables, wood to drive steam engines and so on, and supplying and carting this equipment was a major undertaking. Thus, within a short time bores and drilling became an important feature of inland life. Palmer recalls that by 1896:

> The Government have sunk a number of wells, while hundreds of flowing bores that now stud the great western country have been put down by private enterprise. The policy of the Government has been to determine the area within which artesian water may be hopefully searched for, and to provide water in arid country or on stock routes ... About 800 private bores have been sunk in search of artesian water in the western area of Queensland; of these 515 give a total output of 322 millions of gallons in the twenty-four hours, and the total cost of them amounted to nearly two million pounds.[533]

Generally, percussion drilling rigs were used. With this equipment, steam-powered pulleys wind a long, hollow, heavy cylindrical steel-cutting tool up a tower. The clutch is then released, allowing the tool to drop down the steadily deepening bore hole and cut into the soil or rock at the base. The cutting tool is then winched back to the surface and the tightly compressed rock and soil in the centre of the cylinder is cleared. The rate of drilling depends on the type of rock and soil encountered. Number 12 bore on Strathdarr, for instance:

was sunk in 1940, using a percussion drill. The daily log shows the total depth at 3200 feet, and at 2000 feet on some days they only penetrated about 18 inches.[534]

Once a bore was sunk, steel casing was fitted to stabilise the walls. Usually ten inches in diameter for the first two hundred feet, this diminished in size as the hole progressed. From two to six hundred feet, eight-inch pipe was used and five- or six-inch pipe thereafter. In many instances hydraulic pressure forced the water to the surface, often under considerable force, and water turbines were fre-quently fitted to the bore-head to drive generators, shearing shed plants and so forth. Frequently this water was scaldingly hot. The Government bore at Winton, for example, delivered 720,000 gallons per day from a depth of 4010 feet, at 196 degrees Fahrenheit.[535]

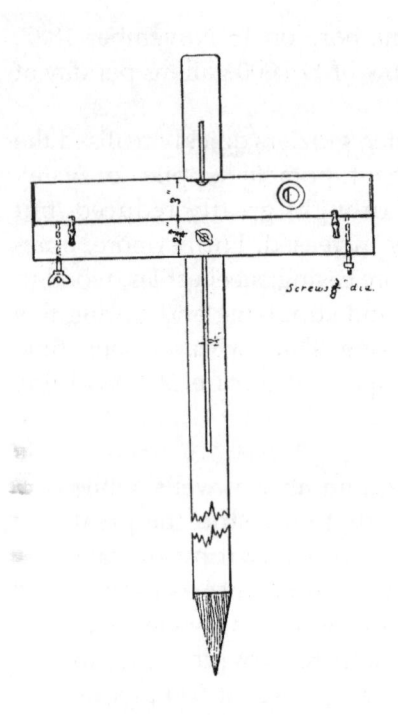

Plate 73: Sighting device with built-in spirit level.

Where water gushed out in huge quantities, bore drains were constructed. These not only carried the water for miles beyond its source, allowing stock to be watered more efficiently, but were considerably cheaper than using piping. When constructing drains it was essential that the route was based on carefully established levels. With the equipment shown in Plate 73:

the flow down these [drains] can be regulated to a nicety if the levels have been carefully taken and the bore site well chosen. The size of the drain depends on the flow.[536]

The selected route was initially deep ploughed to loosen the soil and then the drain was constructed by a horse-drawn implement. These drains usually ran through open paddocks, where they were subject to damage from stock. As poorly maintained drains could lose up to fifteen thousand gallons per mile, repairs were made with clay-rich soil. Stations also constructed channel cleaners, known in Queensland as bore-drain delvers, which were pulled by bullocks or horses (Plate 74). In spite of these precautions, it is calculated that stock consumed only ten per cent of the water flow as a result of soakage and evaporation.[537] Nevertheless, a seemingly never-ending supply of water flowing quietly through miles of drains must have been immensely satisfying to graziers. Paterson expresses this wonderfully in his evocative 1896 poem 'Song of the Artesian Water.'

> It is flowing, ever flowing, in a free, unstinted measure
> From the silent hidden places where the old earth hides her
>   treasure …
> And its clear away the timber, and its let the water run;
> How it glimmers in the shadow, how it flashes in the sun!
> By the silent belts of timber, by the miles of blazing plain
> It is bringing hope and comfort to the thirsty land again …
> To the tortured thirsty cattle, bringing gladness in its going;
> Through the droughty days of summer it is flowing, ever
>   flowing.[538]

Unfortunately, the unceasing flow from these uncapped bores has not been without cost. Within four years of the first bores being drilled it became apparent that local discharge was in many cases exceeding replenishment. Accordingly, in 1891 the

FRONT VIEW.

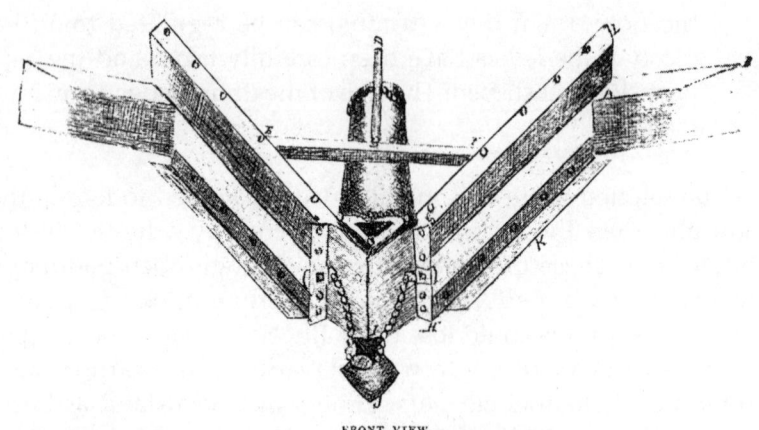

E to F—4 ft.
G to L—4 ft. 6 in.
G to H—24 in.
I to J—8 in.; has to be short and strong.
C to D—3 ft.
A to B—Wings, 24 in. long 16 in. wide.

E to F—6 x 4 timber.
Log—12 ft. x 18 in.
K—Steel plate, 3 in. wide.
Wings can be made of steel plates, 12 in. wide $\frac{3}{8}$ in. thick.

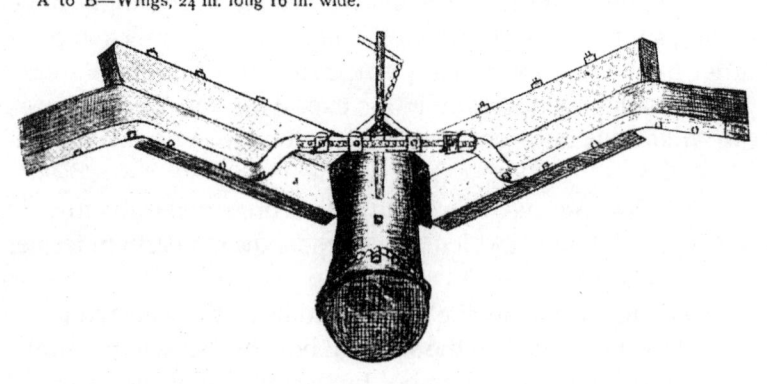

Plate 74: Bore-drain delver.

Queensland Government attempted to introduce legislative control, but the Bill was defeated and not passed until 1910. Similar bills were introduced in New South Wales in 1894, 1897 and 1907 before being passed in 1912. Given this legislative authority, systematic measurement could proceed, and the two governments collated an immense amount of detail regarding every bore sunk in their states. *The 1917 Bore List of Queensland*, for instance, records the driller, where located, date of commencement and completion, depth at which water was reached, total depth, initial flow rate, flow rate after twelve months, and so on.

Subsequently a number of national conferences on artesian water were held. The first of these, in 1912, considered such matters as:

> the limits of the Great Artesian Basin, its intake areas, natural outlets, and the origin and movement of water in the Basin. The problems of diminishing pressure and flow of wells, the corrosion of well casings, and other related matters were also covered.[539]

The overall picture is that in many areas the watertable has declined considerably. In the Julia Creek to Longreach region, for instance, the water level in bores that once flowed freely has dropped sixty to eighty feet. Accordingly, where hydraulic pressure was insufficient to force water to the surface, or where sub-artesian water was being drawn, as occurred in much of the Northern Territory, increasingly sophisticated pumps were applied. According to Dean, with a strong wind the twenty-seven- to thirty-foot head Comet windmills equipped with twelve-inch hot-water pumps could lift 150 gallons per minute. Over twenty-four hours they were capable of lifting 250,000 gallons. Moreover, these mills were self-governing and would not run at more than a set speed, beyond which the main wheel turned side-on to the wind and slowed down.

Steam-driven pumps were also used, especially where wind could not be relied upon. Fuelling these engines required a tremendous amount of wood and unarguably contributed to the deforestation and degradation of large though discrete areas of rangeland. According to former Territory cattleman Felix Schmidt of Alroy Downs:

during my management period firewood was carried from the desert country, up to 20 miles distant ... [and] I noticed a great shortage of shade on the downs and around the bores. Even along the Playford River and creeks all trees had been cut down to feed the steam engines ... It was distressing to see the lack of shade when cattle came in to drink in summer, and calves dying and trying to get under the shade of the trough.[540]

Adequate fencing was also vitally important in rangeland management. Without it, cattlemen had no control over their stock and overgrazed areas could not be rested. Unfortunately, as mentioned in the chapter 'Mustering', although fencing was common throughout the Longreach region 'well before the close selections were taken up, which started in Longreach in 1889', and other regions of Queensland,[541] it was not an everyday feature of Kimberley and Territory stations until the 1960s.

Equipment and Know-how

Whereas the preceding chapters have dealt with broad topics, it has to be recognised that everyday life in the northern cattle regions also demanded proficiency in numerous small ways. Queensland saddler, historian and author Ron Edwards has dealt with this subject comprehensively, having published over forty well-illustrated books such as *Counterlining Stock Saddles, Bridles Plaited and Plain* and *Bushcraft* (vols 1 to 8). Accordingly, rather than reinvent the wheel, this final chapter deals briefly with some of the more interesting of these innovations and modifications that emerged in these areas.

Like the earliest explorers, who quickly found that extended bush life was extremely hard on clothing, the overlanders and early stockmen soon modified theirs to suit conditions. Riding boots showed specialised development. Ideally, these should be as smooth and free as possible from rough surfaces, and high enough to ensure that, in the event of a rider being thrown, his foot will not catch in the stirrup. For this reason lace-up shoes are automatically excluded. Furthermore, bushmen's boots had to be substantially more robust than those used by town dwellers. On Lieutenant Barrallier's 1802 exploration of the Blue Mountains, the standard issue British Army boots gave way and most of the expedition members sustained lacerated feet. Not until 1826, when convicts manufactured '94 pairs of boots for the Mounted Police and Mr Cunningham's exploring party,' was footwear constructed specifically for the conditions encountered in extended periods of bush life.[542]

Given their military origins, these would have been knee-length leather pull-on boots, and thus it is not surprising that in the late 1800s La Meslee refers to early stockmen wearing soft wellington boots and Palmer speaks of cossack boots. Yet, within a little over three decades, Paterson records the remount horsebreakers wearing:

> elastic-sided [ankle-length] boots specially made in Australia, with smooth tops so there would be nothing to catch a rider's foot in the stirrup.[543]

Elastic-sided boots have several advantages over knee-length boots, being lighter, not as hot to wear and considerably cheaper to manufacture. Furthermore, because of the elastic inserts, they are tighter fitting, so providing greater security and comfort. Although an exact date for the transition from knee-length boots to ankle-length elastic-sided boots is not known, Palmer's 1890s mention of stockmen wearing leggings provides a clue, for leggings are worn only with shorter boots.

Later refinements to elastic-sided boots included hidden seams at the back to prevent spurs from rubbing the stitching away. Strips of leather were also stitched vertically across the middle of the elastic to protect it from being chafed by the stirrup leather. A high, backward-sloping cuban heel was often incorporated to prevent a rider's foot slipping through the stirrup should he come off.

In the US, cowboys' boots also gradually diminished in

height. Mora saw this as a retrogressive step, and attributed the change to what Jim Hoy calls the romanticising and mythologising of the West.

> With the passing of the old-timers, the moving picture heroes commenced designing what a cowboy should wear, and that's when they started getting fancier and fancier, and lower and lower till they got to the modern 'peewee' which is quite worthless in my opinion.
> I'm just old-fashioned enough to still consider a cowboy as a working man and his boots according. Well, you can't tuck your pants into one of these contraptions because they just won't stay in; and you can't wear them outside because they won't stay out. They will catch on the edges and appear like they are trying hard to get back in, and altogether they look like hell.
> Then again, have you ever been a flanker at a branding, 'rasslin' calves? Well, if you have you know that you are sitting, or kneeling, or wiggling around on that dirty, dusty ground about as much as you are operating up on your hind legs, and if you're unlucky enough to be wearing a pair of those jokes, with tops about 8 or 9 inches high, you'll no doubt unhappily remember that every rock, pebble and all the loose dirt within the radius of your activities ... hopped over and into [your boots] ... And if you're riding through the chaparral, your overalls are bound to work up over their tops, and it's mighty little brush you'll miss packing home with you inside those so-called boots.[544]

Curiously, although Mora also devotes some pages to the various forms of chaps, which were used for protection from brushwood and mesquite as well as to help keep warm in the winter, and how most cowboys used leather gloves, neither article was common in Australia. Until the advent of broncoing, very little roping was done and gloves were not necessary. Even then, because the end of the bronco rope was fixed to a breastplate, ropers did not have to hold the rope once they had lassoed a beast.[545] To some extent the hot climate also told against leather chaps. Leggings, however, have been a feature of the Australian stockman's clothing for at least a century. They are worn to prevent the rider's instep rubbing against the top of the stirrup, and for protection from grass seeds. Morover, they provide a smooth outer surface where lace-up boots are worn, thus reducing the likelihood of the rider's foot being caught should he come off. The Australian light-horsemen wore strap-on leggings over their standard issue lace-up military boots.

A variety of leggings have been used, including the simple wrap-around type worn by the light-horsemen. These are fixed with two buckles at the top, and a long strap which wraps twice in a spiral around the wearer's leg before being buckled. Later, leggings made of heavy leather with several concertina folds in the middle became very popular. Due to the extra amount of leather and additional work involved in their construction, concertina leggings were gradually replaced by plain leather leggings. These have a flat six-inch vertical insert of spring steel on the outside to give them greater rigidity. In turn, these were replaced by plain pull-on leggings which, having no steel inserts, are lighter and can be made to any length.

Curing and tanning leather were extremely important on stations, not only to have the material on hand for repairing saddles and constructing hobble straps, bronco ropes and stockwhips, but also for making beds, furniture and so forth. Greenhide, which is untanned leather, was most commonly used. To produce greenhide, a freshly skinned hide is firstly scraped free of fat and flesh, then salted and left folded for some days. It is then pegged out on boards, or stretched across a wagon wheel, to dry. These greenish coloured hides are then stored for some months, during which time they set like mild steel plate. Indeed Leichhardt, who was ill-equipped with containers for his scientific samples, wrapped fresh hides around the saddle-bags containing his invaluable botanical specimens and allowed them to dry *in situ*. Just before a dry hide is to be used, it is put:

> in fresh water, say at noon ... Take it out of the water again at dusk, and place it in a room or shed where no wind can blow upon it. Spread it out upon the floor flesh side down, and leave until the following morning, when the water will have dried off and the hide [will] be found of an even consistency, somewhat resembling new cheese.[546]

In this state, the greenhide is ready for cutting. Beginning either in the centre or at the outside of the hide, a long continuous strip is pared off with a sharp knife. Strips vary in width, depending upon their intended use. Ropes, for instance, require strips about an inch wide. *The Stockowner's Guide* advised attaching one end of the strand to:

> a strong post, pulling the opposite ends out full length from the post, then securing two or three [station] hands to try and break the hide by pulling it. If the strand breaks, throw it away and commence again with a fresh hide.[547]

When making bronco ropes, ideally the straps were bevelled and shaved, so a uniform piece of only the best leather remained. Often, however, the hair was not even removed and it was not unusual for greenhide ropes to vary markedly in quality and appearance. Whatever the approach, the single long strand was then cut into three pieces of equal length and attached to a rope-making machine. These ranged from simple planks attached to a wagon wheel, to specialised contrivances. In all cases the principle was the same. Three parallel strands were attached to separate crank handles on a header bar, and tied at the other end to a heavy weight, such as a two-hundred-pound anvil or a large bag of sand. The strands were then individually turned by their crank-handles until they had wound up and begun to buckle. The operator then went to the other end and began winding the three strands together in the opposite direction.[548] As the rope was gradually formed, it shortened and dragged the weight forward, thus maintaining the necessary tension. The ends were then tied off and the ropes liberally coated with grease or oil. Although stiff to begin with, greenhide ropes quickly become supple with use.

Greenhide was also used for beds and chairs. After being nailed or laced into place, the wet greenhide shrank as it dried, thus providing a very firm base to a bed or seat. Even today some saddlers use greenhide to bind trees together, the finished article being sufficiently smooth and strong to be mistaken for fibreglass.

Although stockwhips were initially often made of greenhide, it was soon found they fell better if made of tanned leather. Tanned leather was usually bought in as required, though some tanning was carried out on stations. Alum was applied with the salt, or a mixture of neatsfoot oil and brains rubbed in the pegged-out skin every day for some weeks. Alternatively, wattle bark was covered with boiling water and hides immersed in the infusion for some weeks.

While kangaroo hide unarguably produces the better whips, redhide was commonly used, even though 'the old died-in-the-wool whipmakers said [it was] ... absolute rubbish.' Certainly redhide whips do not fall as well. Nevertheless, as third-generation whipmaker Brian Nemath points out:

> stockmen in Australia have used hundreds of thousands of them ... I've made [many thousands of] whips and the bulk of them are redhide. They don't work as well as roo-hide, but they are a good lasting whip.[549]

Whatever leather was used, the approach was the same. The hide was firstly cut into thongs, working:

> from the outside in. Reg Williams cuts from the centre [outwards], but with this method you only get one whip out of a roo hide, no matter how big. Quite often you mightn't even get a whip if you started in the wrong place.[550]

Then, by using wooden gauges, each thong was subdivided into four strands which were tapered towards the tail of the whip.

Most kangaroo-hide whips were made in eight, ten or twelve plaits, and had a belly (centre piece) of rolled, split leather.

> Some use football or embossing leather. We cut, edge and shave it, then … bevel the outside edge with a spokeshave. It's an old wooden spokeshave, which was for making wooden wheels and suchlike. They aren't made anymore. They all originally came from England … Then we wet it and fat it [the leather], then roll it. We plait over that for about three feet, which is called a plaited belly whip.

Unfortunately, due to the economic realities of contemporary life, plaited belly whips have become a rarity. The best whips were also lead-lined, to improve their fall. Nemath still uses the principle, albeit with some modification.

> In the old times they used to start probably from about a foot from the keeper and [put lead in] down another two feet, in a seven foot whip. You'd run lead shot [from a shotgun cartridge] into a little muslin bag. It's a slow, laborious and costly way of doing it and nowadays we use a piece of cable, such as speedo cable. It gives it the weight and flexibility. It makes the whip fall better, that's what it's done for. On the

handle we tack on sheet lead and cover it over, then plait over it, which gives balance in the handle.[551]

Once hobbling became standard practice, and animals could wander several miles through heavy scrub overnight, bells were rapidly improved and refined. The earliest bells were usually flat-sided and wider at the base than at the top. They were not efficient, for the clapper tended to lie along the bottom edge of the bell and would not ring unless the animal shook the bell with some vigour. Probably the most important innovation was the development of the Condamine bell. These bulge at the centre, then taper in again towards the mouth. As a result, the clapper is not resting full-length against a flat surface and rings more readily.

Welsh migrant Samuel Jones of Condamine devised these bells, constructing the first one from the high carbon steel of an old pit-saw blade. This type of steel was used because it had been found to resonate longer and more strongly when struck, and thus the sound carried further. With experience, Jones became increasingly demanding over the materials and construction of his bells. He used:

> only charcoal to fire his bellow-blown forges and he was most particular about its quality. He would not use charcoal which was burned in a pit, this being the normal method of producing it. Charcoal from ironbark was of no use as it threw too many sparks, and charcoal from spotted gum was also barred. Grey gum was, however, acceptable.[552]

Jones' insistence on specific high-quality materials is understandable, for bells indisputably vary in tone and carrying range. According to Edwin Bischoff of Wandoan, son of the renowned scrub-musterer and bullock teamster Stan Bischoff, the simplest and most accurate method of checking a bell's carrying potential is to ring it and hold a hat above it. The greater the resonance, the more the hat will vibrate and, by moving the hat further and further away, some indication can be gauged of the bell's resonant qualities and carrying range.

Thereafter, as orders increased, purchasers had to supply Jones with the necessary saw blades. Eventually:

> the demand exceeded the supply and so Samuel Jones sent to Brisbane for sheets of English Netherton iron, both heavy gauge and lighter gauge, approximately 72 in. by 30 in. There was a total of 16 graduated sizes. Camel drivers and bullockies generally required larger bells than could be made from most saws.[553]

Unarguably an element of one-upmanship arose, with drovers and stockmen often boasting of the size of their bells and how far they could be heard. Some stories strained credulity, however, especially claims of the sound carrying for twelve miles through heavy forest country. Conversely, very small bells, known as tinklers, were used to counteract the tendency of most animals to rest quietly in the last hour or so before dawn, when horse-tailers and teamsters were searching for them in the dark. Tinklers rang out from even the smallest movement and thus were the 'give-away' bells.

Bells had drawbacks, however, for during the day, as the horse plant was being driven along, the bells rang incessantly. To overcome this, a strip of leather was fitted over the clapper to hold it in the centre of the bell. At night, before the horses were hobbled out and released, the leather strip was removed, thus allowing the clapper to move freely again. Moreover, should the belled horses stay close to the camp, especially when strange animals were causing fights and upsetting the team, sleep could be disrupted. Searcy captures the annoyance bells could cause in his humorous account of life on the outskirts of Darwin in the 1890s.

> In Darwin, nearly all the residents who had horses turned them adrift at night, belled, of course … As long as the animals kept a little distance from the camp

[house] all was well — in fact, then the sound of the deep-toned bell is rather pleasant ... [but] These animals always seemed to make for my quarters ... [and] to be suddenly awakened by the infernal jangle of bells, caused by the animal's frantic endeavours to get at some part of its body (which it was quite impossible to do) where the insect pests had decided to have a little fun, was calculated to cause a flow of language more or less tropical ... so I was nearly driven mad by the confounded din. I tried my hand at reducing the nuisance. The horses were driven into a yard and all the bells removed and mysteriously lost in the bush. People who at first could not imagine what had become of their bells at last grew suspicious. Chains and padlocks were then used instead of leather straps. I must confess that I rose to the occasion. With a powerful pair of pincers the tongues were wrenched out. The horses appeared to enjoy the fun, but the people did not. It just about killed the nuisance.[554]

A number of other quite simple items also underwent development. Stockmen, for instance, needed to carry a small billy and mug, so they could quickly boil up a cup of tea at lunchtime. Conventional billies were too bulky and difficult to attach to the saddle, and thus the ubiquitous quart-pot was developed. This is a small metal billy of one quart capacity and a cup which, when not being used, fits tightly inside its open end. Both are then placed inside a leather carrying bag strapped to the saddle. Being either oval or flat-sided, they fit more closely against the horse and are less vulnerable when galloping through dense scrub.

Before pack-saddles became commonplace, carrying equipment was not always easy. The first saddle-bags were a jute sack with the top sewn closed and one side split laterally across the middle. Known as split-bags, they were thrown across the horse's loins and the ample pockets filled with

Plate 75: Early saddle-bags.

clothes, food and so on. Later, more substantial bags with buckle-down flaps were constructed of canvas or leather, as shown in Plate 75. With the advent of pack-saddles and wagons, stockmen did not need to carry as much on their riding horses, and saddle-bags became smaller. Thus the present-day article is less than twelve inches by eight, and is a single bag. By contrast, early US cowboys overcame the problem of saddle-bags by constructing them as an integral part of the saddle, along Mexican lines. These made the saddle exceedingly cumbersome and heavy, and according to Mora, Texan saddlers soon discarded them. In fact saddle-bags were never popular with cowboys who, if they had to carry anything extra, would usually:

> roll it in a gunny sack or into the folds of their slicker, [then place it] snug up close to the back of the cantle and tie it down with the saddle strings.[555]

Spurs were often necessary and a very wide range have been made in Australia. Campbell always used them.

> On several occasions they got me out of trouble on tired horses. I had lots of near misses and if it wasn't for these gentle persuaders, I would not have got through.[556]

Traditionally they resembled British hacking or military spurs, which had a short neck, small or blunt rowels and a strap that wrapped completely around the wearer's foot. Spurs of this type were not particularly suitable for Australian stockmen, who rode with their feet much further forward, and a spur with a much longer neck was developed.

Cooking equipment also underwent changes. Early explorers and bushmen commonly used utensils which, being based on British military equipment, were made of cast iron and excessively heavy. Sturt's 1828 expedition manifest, for example, records 'three iron boilers, two long-handled frying pans and an iron kettle.'[557] Another drawback was that they were readily broken by a packhorse brushing against a tree, or by being dropped. As no spares were available in the bush, cooking pots made of steel were introduced. Known as Bedourie ovens, they were reputedly developed by Pompey Trew of Clifton Hills station, near Bedourie in south-east

Queensland. They are almost unbreakable and, even should a packhorse dash them against a tree, judicious panel-beating sets them right again. Furthermore, not having legs, they are easier to pack.

Being able to keep meat fresh and free from ants and flies was vitally important in stock camps, where there were no Coolgardie safes, let alone refrigeration. Frequently Edgar found that:

> in the wet I'd kill today and tomorrow it'd be blown. I used to leave it in salt, under leaves, all kind of things, but blowflies have got a one-track mind.[558]

As a result, fresh meat was hung immediately in the shade and covered in a carefully tied calico or hessian meat bag. It was then cooked as soon as possible. Alternatively, the beast was cut down and the meat sun-dried or salted. Salting entails making deep cuts in each piece of meat and liberally rubbing in coarse salt. The salt draws moisture from the meat until it is quite desiccated. During the curing process, the meat is turned daily and re-salted where necessary. When salt-dried meat is to be used, a number of long, dark-coloured, seemingly unappetising pieces are taken from the store room and boiled once or twice to remove the salt. The surprisingly flavoursome meat is thus rehydrated.

Leichhardt was the first explorer to sun-dry meat. He took eight pack bullocks and eight for slaughter, which he killed and dried as required. The process involved cutting the meat into strips and hanging them from branches of trees and shrubs around the camp. Depending on the weather, one to three days were spent continuously turning the meat to expose it to the sun and in keeping predatory kites (hawks) away. Leichhardt refers to this technique in his *Journal*, pointing out that he dried the meat according to the method the South Americans use to prepare charqui. Apparently he was informed of the practice by employees of the Australian Agricultural Company, which had contacts in South America, having imported mules and horses from Chile. During this process the meat, after being hung outside for:

> an hour or two, in the furnace-hot air of a summer day ... got a glazed surface, which made it impervious to rot, flies, or ants. Flies did not attack in the middle of the day. It was too hot and ants would be unlikely to find the meat until it had dried. But there could be a couple of catches to this drying method. Fat meat tended to go bad before the surface hardened. If there was a chance of flies settling near, they were almost sure to blow the meat along the bone, as it remained slightly moist in that quarter. So it was better to have lean, boneless meat for drying purposes.
>
> Properly dried meat would keep indefinitely. It soon got a glazed skin.[559]

Finally, keeping tobacco moist in the dry season and matches dry in the intense humidity of the wet season taxed the ingenuity of even the most ardent smokers. As usual, Searcy provides a humorous outline of their travails:

> How often we smokers blessed the [dry-season] south-easters, for they dried the tobacco so that the action of rubbing it turned it into dust. In the wet season it is just the opposite; the trouble was to keep the tobacco dry. Many and many a time I had to dry mine over the lamp before I could get a decent smoke. The wet season is very rough on matches. We nearly always used wax ones. What trouble we had to keep them at striking pitch. During the wet I always found blotting-paper the best to strike them on. Having to hold the match close to its head, it was no uncommon occurrence for a piece to lodge under your thumb

when it ignited. In all my experience I never realised anything so calculated to cause extraordinary antics and language as a piece of wax vesta burning under your nail.[560]

Epilogue

Few teachers and students of Australian history, particularly early bush history, would deny that their response to the subject is one of 'collective groans,' if not antipathy. Indeed, Professor Geoffrey Bolton stated in his 1996 keynote address to the History Teachers' Association of Western Australia that as a child, when:

> tracking the big boots of explorers as they trudged their way up some dry gully or other [I found the subject] ... very remote from the here and now of growing up in a suburb in Perth or Melbourne.[561]

This outlook upsets me deeply, for bush history is not inherently lacking in interest. It seems the primary reason for this response is that between 1947 and 1986 the rural component of the Australian population declined from 31.1 to 14.5 per cent.[562] Thus, whereas baby boomers frequently had a family member or close relative engaged in farming or pastoral pursuits, with whom they often spent some time as children, such is not the case with Australian children of the 1970s and later. The urban–rural link has gone, and while images of bucking horses, cattle rushes and vast distances are still richly evocative for many rural and older Australians, they are now irrelevant to most urban and younger people.

Moreover, the subject has not been academically fashionable for the past thirty years or more, having been supplanted by newer, more theoretical concerns. Indeed, it is almost treated with disdain. Interested postgraduates have difficulty in finding supervisors and researchers' funding applications are not viewed favourably, thus affecting the quality and quantity of work in this area. As a result it is hardly surprising that the majority of teachers and university lecturers lack any knowledge or practical experience in this field, and quite possibly any particular interest. It is also the case that, with the exception of Winton and the Stockman's Hall of Fame, rural museums from Longreach to Derby lack funding and professional expertise. Indeed, as commentator and critic Donald Horne patronisingly but accurately pointed out in January 1992:

> we really must no longer encourage country people to degrade themselves by putting together junk heaps of old sewing machines, flat irons, agricultural machinery and dentists' chairs as if these were museums.[563]

Unarguably, in order to attract young Australians and interest them in our early bush history, books will need to be not only authoritative and informative, but also lucid. Furthermore, exhibitions designed to illustrate and display this aspect of Australian history will need to be assembled thematically, and both explanatory and interpretive information provided.

Perhaps the best and worst examples of this notion can be seen at the Riversleigh Fossils Interpretive Centre at Mount Isa, and in the display of twenty or so samples of barbed wire at the Hall of Fame. Barbed-wire fencing revolutionised the cattle industry on the United States plains and to a lesser extent in pastoral Australia, by preventing cattle from wandering unchecked and thus facilitating mustering and herd upgrading. As well, in conjunction with the provision of bores and multiple watering points, soil erosion and degradation could at last be controlled. Yet none of these broader implications of the barbed-wire exhibit are touched on. Like so many other exhibits in northern Australian museums, it quite likely has only minimal interest for farmers and graziers, who have seen numerous other examples in their everyday work, and presumably passes over the heads of urban Australians.

By contrast, the Riversleigh museum incorporates professional audiovisual and lighting displays in conjunction with superbly crafted models of animals based on fossil remains. These models are placed in simulations of the creatures' normal environments, which include running streams. Professionally illustrated picture boards are also used in conjunction with fossil remains, or model replicas. Aspects developed include the origins of the animals, why the Riversleigh fossils have been so well preserved, what the country looked like then and how it appears now, the means and difficulties of field research on site, the processing and interpreting of samples in laboratories, and the relationship of these beasts to modern-day animals. Overall, the display is explanatory and sequentially coherent, with each section amply illustrated by a variety of media. More importantly it is, as its title proclaims, an interpretive centre, where broader interlinking questions are posed and answered.

Unfortunately displays of this calibre are extremely costly to establish and run. Riversleigh's list of sponsors includes Ansett Australia, the Commonwealth Department of Tourism, the Mount Isa Chamber of Commerce, the Queensland Department of Environment and Heritage, the Queensland Museum and Mount Isa Mines Limited. It seems that unless northern rural museums can obtain similar funding and professional assistance, and until practically experienced and academically qualified historians are encouraged to write on this subject, there is little likelihood of this fascinating aspect of our cultural heritage gaining due recognition in the Australian community.

Accordingly, as my own small contribution to a subject which I believe in deeply, I have three wishes for this book. Firstly, that everyday Australians — urban and rural, men and women, young and old — will read and enjoy it. Secondly, that students of cultural heritage studies will find it an informative and interesting introduction to the subject. Finally, that teachers, lecturers and museum curators will draw from it to help bring alive the pragmatic workaday achievements of these remarkable pioneers.

Notes

Introduction

1. Jim Hoy, 'The Americanisation of the Outback: Cowboys and stockmen,' unpublished paper, 1995, p 9. Hoy is explicitly clear, however, that his opinion is based only on 'impressionistic responses to conversations and observations,' and definitely not the findings of 'any sort of scientific study.'
2. ibid., pp 3–8.
3. ibid., p 11.
4. ibid., p 11.
5. The northern pastoral cattle industry embraces all cattle run under pastoral conditions. Obviously this area has varied over time, as clearing in the south of Australia has extended the boundaries of farmland into what previously were pastoral regions. Overall, it can be said that the pastoral cattle industry extends from the Murchison in Western Australia to central Queensland.
6. H M Barker, *Droving Days*, Hesperian Press, Perth, 1994, pp 112–3. See also Keith Willey, *Boss Drover*, Rigby, Sydney, 1971, p 162.
7. Brucellosis and Tuberculosis Eradication Campaign.

The Beginnings

8. Phillip to Sydney, 9 July 1788, *HRN*, vol 1, part 2, p 149. By contrast, the British Museum Papers, Appendix E, 21 December 1795 (*HRN*, vol 2, p 820), note that one bull and six cows strayed.
9. George Boxall, *The Story of the Australian Bushranger* (facsimile edition), Penguin Books Australia, Melbourne, 1974, p 45.
10. British Museum Papers, p 820.
11. ibid.
12. 'A letter from Sydney,' 31 October 1799, *HRN*, vol 3, p 730. Of these cattle, Boxall comments (p 45) that 'These animals are much more active than the

fine-boned, heavy-bodied, short-horned, or other fine breeds, but they can never be properly tamed.'

13 King to Portland, 10 March 1801, *HRN*, vol 4, pp 321–2.
14 King to Portland, 14 November 1801, *HRN*, vol 4, p 622.
15 King to Hobart, 1 March 1804, *HRN*, vol 5, pp 319–20.
16 Governor King's remarks on 'George Caley's observations on the Cow Pastures,' 2 November 1805, *HRN*, vol 5, p 719.
17 Bligh to Windham, 31 October 1807, *HRN*, vol 6, p 358.
18 Macquarie to Liverpool, 18 October 1811, *HRN*, vol 7, p 601.
19 By late 1839 at least 7088 horses, 371,699 horned cattle and 1,334,593 sheep were spread over 694 licensed properties in New South Wales, covering an area 900 miles long by 300 miles deep. Gipps to Russell, 28 September 1840, *HRA*, S1, vol 20, p 839. Gipps noted ironically, however, that as pasturage fees were paid on livestock numbers, these were almost certainly an underestimate.
20 The widely experienced stockman and camel teamster H M Barker, in *Droving Days*, Hesperian Press, Perth, 1994, pp 6–12, provides a brief analysis of Joseph Hawdon's pioneering feats as a drover. Hawdon narrowly beat Eyre to Adelaide with the first stock overlanded from Port Phillip and, in a career lasting only about six years, undertook several other important journeys. Consequently Barker considers 'he had more influence on the future of droving than anyone before or since his time.'
21 Edward John Eyre, *Autobiographical Narrative of Residence and Exploration in Australia 1832–1839 by Edward John Eyre*, ed J Waterhouse, Caliban Books, London, 1984, pp 75–6.
22 ibid., p 190.
23 Hawdon's exploits are outlined in the chapter 'Droving'.
24 George Grey, *Journals of Two Expeditions of Discovery in North-West and Western Australia* (facsimile edition), vol 2, Hesperian Press, Perth, 1983, pp 184, 188–9.
25 Edward Palmer, *Early Days in North Queensland* Angus and Robertson, Sydney, 1983, pp 85–6.
26 ibid., pp 34, 39. While technically correct, this assessment fails to give due weight to William Landsborough's achievements. On returning in 1862 from his arduous and unremitting, though unsuccessful, attempts to find some trace of Burke and Wills, Landsborough brought valuable information on the topography of the Barkly Tableland and the central and upper Flinders basin. These were all areas of superb and extensive natural grasslands, eminently suitable for pastoral purposes.
27 ibid., p 111.
28 Noel Loos, *Invasion and Resistance* Australian National University Press, Canberra, 1982, p 29. Citation details provided in endnotes 5 and 6, chapter 2, p 256.
29 Barker, pp 28–30.
30 Palmer, p 101.
31 Nehemiah Bartley, *Australian Pioneers and Reminiscences, 1849–1894* (facsimile

edition), John Ferguson, Sydney, 1978, pp 367–8. See also Barker, pp 23, 26–7.
32 Palmer, p 128.
33 M. Powell, The rise, courses and consequences of the crisis of 1866 in Queensland, BEc thesis, 1969, p 103.
34 Bartley, p 196.
35 Palmer, p 141. Palmer also notes that the lifting of the financial depression helped greatly. Wool, for instance, doubled in value. For a more detailed outline of the development of mining in northern Queensland and its effect on cattle prices and the reinvigoration of the pastoral industry, see Palmer, pp 136–142, and Loos, pp 62–6.
36 *Official Year Book of the Commonwealth of Australia*, no. 1, Commonwealth Bureau of Census and Statistics, Canberra, 1908, pp 279, 284.
37 Unthrifty country is soil on which stock do not thrive.
38 Loos, pp 51–3, covers the question of economic disadvantages in dangerous regions. Barker, p 30, notes that George Sutherland described the Barkly Tableland as 'second rate cattle country and third rate for sheep,' yet points out that horses thrive there.
39 Palmer, p 129.
40 Barker, p 31. While Barker's claim has a romantic ring to it, in reality a number of eastern Tableland properties ran sheep very successfully. Alroy Downs, for instance, was running 60,000 head in 1913.
41 Ion Idriess, *Over the Range*, Angus and Robertson, Sydney, 1942, pp 215–18.

Mustering

42 Very few stations in the Northern Territory and Kimberleys had bullock-paddocks in which to hold the selected fat bullocks. Generally, the steadily growing mob was held under control, day and night, by stockmen.
43 Reg Wilson, Battye Library accession no. OH 2730/12, Question (Q) 26.
44 Wilson's father subsequently took out a butchering licence at Brock's Creek, 104 miles south of Darwin on the Palmerston– Pine Creek railway line and also supplied fat bullocks to Darwin butchers during and after the Depression. Consequently, although he did not want to muster in the wet again, he had no option.
45 Jock Makin, *The Big Run*, Lansdowne Press, Sydney, 1996, p 121. This was almost certainly an overestimate.
46 *Official Year Book of the Commonwealth of Australia*, no. 46, Commonwealth Bureau of Census and Statistics, Canberra, 1960, p 959.
47 *Northern Territory Times*, 23 March 1906.
48 *Year Book*, no. 23, 1930, pp 86, 456.
49 In the wet season, with plentiful feed, cattle congregate in larger mobs. Dry season foraging groups are much smaller.
50 Ion Idriess, *Over the Range*, Angus and Robertson, Sydney, 1942, p 43.

51 Walkabout disease is a fatal complaint caused by ingestion of *Crotalaria* species. See the chapter 'Horsemanship' for further details.
52 Jo Mora, *Trail Dust and Saddle Leather* University of Nebraska Press, Lincoln and London, 1987, pp 27–8. Mora also outlines the initial dissension caused by the introduction of barbed wire.
53 This difference was almost certainly due to the better returns achieved by US ranchers. See the chapter 'Droving' for further details.
54 Chas A Siringo, *A Texas Cowboy* (facsimile edition), Time Life Books, Virginia, 1980, pp 159–60.
55 ibid., p 167.
56 ibid., p 168.
57 Keith Willey, *Boss Drover*, Rigby Limited, Sydney, 1971, p 112.
58 ibid.
59 Mrs Aeneas Gunn, *We of the Never-Never*, Hutchinson, London, 1962, p 168.
60 ibid., p 174. Interestingly, on p 169 Gunn claims that station Aborigines were often reluctant to divulge the location of many permanent waterholes to early pastoralists, lest one day they need a refuge.
61 The Emmanuels were a wealthy Western Australian family with extensive pastoral interests in the Kimberleys.
62 Dave Ledger, Battye Library accession no. OH 2730/7, Q10 and Q12.
63 Wilson, Q2. See also Glen Edgar, Battye Library accession no. OH 2730/4, Q20.
64 Wilson, Q8.
65 ibid., Q8. See also Edgar, Q21.
66 Thomas Cockburn-Campbell, *Land of Lots of Time*, Fremantle Arts Centre Press, Fremantle, 1985, p 113.
67 Mora, p 172.
68 Ledger, Q14. See also Bruce Simpson, Battye Library accession no. OH 2730/10, Q3.
69 Simpson, Q3.
70 In dense scrub and vine country, large numbers of cattle were rarely mustered at one time. According to Robert Bradshaw, when mustering the rugged Glenhaughton station west of Taroom, 'four would be very good.'
71 Stan Bischoff, 'Wild cattle, scrub and prickly pear,' ed Jennifer Bischoff, unpublished memoirs, p 20.
72 Wilson, Q33.
73 Charlie Schultz, *Beyond the Big Run*, ed D. Lewis, University of Queensland Press, Brisbane, 1995, p 102.
74 Mora, p 132.
75 Ledger, Q15.
76 ibid.
77 Schultz, p 80.
78 These straps frequently had metal keepers on the buckles, rather than leather, for a thrashing bullock could cut leather keepers on a stone and get free.

79   Simpson, Q3.
80   Bischoff, p 83.
81   Neale Stuart, *The Last of the Early Day Scrub Riders: Stan Bischoff*, B.I.G. Printing Services, Toowoomba, p 38.
82   ibid., p 42.
83   That is, not try to steer it.
84   Bischoff, pp 28–9.
85   Stuart, p 60.
86   ibid., p 61.
87   Bischoff, pp 57–9. The horse walked some miles back to its paddock overnight and was back at work mustering a few months later.
88   ibid., p 59.
89   Stuart, p 56.
90   ibid., p 39.
91   Bischoff, p 39.
92   Stuart, p 40.
93   ibid., p 14.
94   H M Barker, *Droving Days*, Hesperian Press, Perth, 1994, pp 43–5.
95   Bischoff, p 19.
96   Nehemiah Bartley, *Australian Pioneers and Reminiscences 1849–1894* (facsimile edition), John Ferguson, Sydney, 1978, pp 200–1.
97   Gunn, pp 189–90
98   Bischoff, p 56.
99   Reg Wilson of Manton River station recalls that he and his older brother went on their first stock camp muster when he was six. When older and more experienced, he was required to drive the dinner-pack and canteen horses, together with the spare (fresh) riding horses for each musterer.
100   Tom Cole, *Hell West and Crooked*, Collins Publishers Australia, Sydney, 1988, pp 36–7.
101   Simpson, Q2.
102   Schultz, p 95.
103   Mora, p 139.
104   Long poles were too unwieldy to use, and the following reference to Sam Buckley carrying an eight-foot pole needs to be viewed with some scepticism.
105   Henry Kingsley, *The Recollections of Geoffry Hamlyn*, ed J S D Mellick, University of Queensland Press, Brisbane, 1982, p 254.
106   ibid.
107   Gunn, pp 188–9.
108   A breastplate wraps around a horse's chest and is held in position on either side by the saddle girth strap.
109   Schultz, p 63.
110   J H Kelly, *The Beef-Cattle Industry in the Leichhardt–Gilbert Region of Queensland*, Bureau of Agricultural Economics, Canberra, 1959, pp 76–7.
111   Mora, pp 186–7.

112 A side-line is a restraining harness of chain and leather, which connects both the front and back left or right legs and thus prevents the horse moving quickly. See the chapter 'Horsebreaking' for further details.
113 Edgar, Q20 and Q56. See also Gunn, p 203, for her description of bogs on Elsey station.
114 Siringo, p 82. See also Mora, pp 135–6.
115 Bischoff, p 41.

Droving

116 A B Paterson, *Singer of the Bush*, Lansdowne Press, Sydney, 1985, p 105.
117 In this light, numerous former drovers have accused Paterson of poetic licence.
118 Edward Palmer, *Early Days in North Queensland* (facsimile edition), Angus and Robertson, Sydney, 1983, p 199.
119 Thomas Cockburn-Campbell, *Land of Lots of Time*, Fremantle Arts Centre Press, Fremantle, 1985, p 126. See also Chas A Siringo, *A Texas Cowboy* (facsimile edition), Time-Life Books Inc, Virginia, 1980, p 126, and Glen Edgar, Battye Library accession no. OH 2730/4, Q17 and Q18.
120 The dust encountered while working in yards was far worse.
121 J T Adams (editor in chief), *Dictionary of American History*, vol 4, Oxford University Press, London, 1940, p 310.
122 R A Billington, *Westward Expansion*, 2nd ed, The Macmillan Company, New York, 1960, p 674.
123 According to Jim Hoy (*Cowboys and Kansas*, University of Oklahoma Press, Norman, 1995, p 141):
> Before the war Texans drove their cattle to ports such as Galveston, from which they were shipped to New Orleans and other markets. They also drove them overland to Missouri, Illinois, and other eastern states, as well as westward to the goldfields of California. These early drives, however, tended to be small and the markets somewhat uncertain.
124 Jo Mora, *Trail Dust and Saddle Leather*, University of Nebraska Press, Lincoln, 1987, pp 13–14.
125 Hoy, p 9.
126 Billington, p 675.
127 Mora, p 22.
128 ibid., pp 29, 142.
129 ibid., p 24.
130 Adams, vol 1, p 327
131 Billington, p 685.
132 Hoy, p 144.
133 *Official Year Book of the Commonwealth of Australia*, no. 46, Commonwealth Bureau of Census and Statistics, Canberra, 1960, p 286.

134 Territory routes were gazetted in the 1930s.
135 H M Barker, *Droving Days*, Hesperian Press, Perth, 1994, p 11.
136 Mora notes (p 143) that this was also the case in the US, where smaller mobs became accepted for they were 'easier to handle and more economical in the long run.'
137 When trucks were introduced, the cook would drive to the dinner camp to prepare a meal, while the horse-tailer continued straight to the night camp.
138 Some drovers also specialised in short distance trips over a set route, preferring not to travel too far from their home base.
139 Edgar, Q18.
140 For some nights the distraught cows would attempt to break away and return to where they last saw their calves.
141 Barker, p 53.
142 Dave Ledger, Battye Library accession no. OH 2730/7, Q30.
143 Campbell, p 118.
144 Edgar, Q34.
145 Mrs Aeneas Gunn, *We of the Never-Never*, Hutchinson, London, 1962, p 181.
146 Wilson believes these would have been Ah Kup's men. At that time Ah Kup ran a butchering business at Maranboy, forty-five miles north-west of Elsey station, and a pastoral property on the Wilton River.
147 Gunn, p 181.
148 Ledger, Q24.
149 Campbell, p 118.
150 Barker, p 52.
151 Edgar, Q49.
152 Peter Ross, Battye Library accession no. OH 2730/9, Q5.
153 Charlie Schultz, *Beyond the Big Run*, ed D Lewis, University of Queensland Press, Brisbane, 1996, pp 110–11. Barker also writes well on this subject and on the question of riders being thrown by the nighthorse shying very quickly from a rushing mob.
154 Siringo, pp 194–5.
155 Ludwig Leichhardt, *Journal of an Overland Expedition in Australia* (facsimile edition), Macarthur Press, Sydney, 1847, pp 182–3.
156 This could be due to leaking tanks or troughs, worn-out mills or drovers watering their stock more than once.
157 Bruce Simpson, *Packhorse Drover*, ABC Books, Sydney, 1996, p 80.
158 See Campbell, pp 103–4, for further details of the process.
159 A J Macgeorge, *Where Fortunes Lay*, ed Michael Macgeorge, Hesperian Press, Perth, 1993, p 39.
160 Nehemiah Bartley, *Australian Pioneers and Reminiscences 1849–1894* (facsimile edition), John Ferguson, Sydney, 1978 , pp 197–8.
161 Joseph Furphy, *Such is Life*, Angus and Robertson, Sydney, 1975, p 68.
162 Simpson, p 140. See also Tom Cole, *Hell West and Crooked*, Collins Publishers Australia, Sydney, 1988, p 23.

163 Furphy, p 170
164 Simpson, private correspondence.
165 Edgar, Q57.
166 Furphy, p 345.
167 Ledger, Q68.
168 Simpson, private correspondence. A shin tapper was a set of hobbles attached to one leg only. Should a horse move too quickly in an attempt to avoid capture, the freely swinging hobble chain would deal it a painful blow across the opposite shin and induce it to stand quietly.
169 Cole, p 148.
170 ibid., p 149.
171 Alfred Searcy, *In Australian Tropics* (facsimile edition), Hesperian Press, Perth, 1984, p 99.
172 Mora, p 159.
173 Schultz, p 154. See also Ion Idriess, *Over the Range*, Angus and Robertson, Sydney, 1942, p 9.
174 Barker, pp 76–7.
175 Simpson, pp 78–9.
176 *The Stockowner's Guide*, Pastoralists' Review, Sydney, 1912, p 120.
177 M Ristic and I McIntyre (eds), *Diseases of Cattle in the Tropics*, Martinus Nijhoff, The Hague, 1981, pp 258–67.
178 Bang-tailing involved cutting the hair at the end of the tail off square, so vaccinated beasts were readily identified.
179 Felix Schmidt, 'Memorabilia of "Alroy Downs",' NT, unpublished memoirs, pp 2–3.
180 Barker, p 79.
181 Ledger, Q28.

Horses and Breeding

182 Appendix C, The journal of Lieutenant King, November 1787 to March 1788, *HRN*, vol 2, pp 530, 544.
183 Paterson to Dundas, 21 March 1795, *HRN*, vol 2, p 286.
184 See Hunter to Portland, 18 November 1796, *HRN*, vol 3, p 182, where Hunter responds to criticism over importing bullocks instead of breeders: 'It is certainly true, my Lord, that the importation of oxen into a colony which was wholly without breeding cattle is a measure which cannot well be defended.'
185 Hogan to Portland, 23 September 1797, *HRN*, vol 3, p 300. This figure was actually quoted on a loss basis, the contract being conditional on the carrier getting a guarantee of government freight on the return voyage.
186 This was the seventeen-hand stallion Northumberland, which was reported to have thrown stock 'much larger than any that were in the colony before his arrival and [which] will be found of very essential service to the colony.' The

Duke of Northumberland Papers, August 1806, *HRN*, vol 6, p 182. According to A B Paterson (*Song of the Pen*, Lansdowne Press, Sydney, 1985, p 346), the stallion Northumberland was 'not a blood horse, but was a utility animal, a cream-coloured Yorkshire horse, so his stock were not likely to race, not being intended for that purpose.'

187 Edward M Curr, *Pure Saddle Horses, and How to Breed Them in Australia*, Wilson and MacKinnon, Melbourne, 1863, p 141.
188 The Duke of Northumberland Papers, p 182.
189 The sources for the graphs are the *Historical Records of New South Wales* and the *Historical Records of Australia*, from settlement to 1845. Problems occurred in the compilation of this data, for the *Historical Records of Australia* are not complete for every year of analysis. Government-quoted labourers' wages were chosen to preclude the possibility of variations occurring between different categories of salaried staff such as surveyors, superintendents and foremen.
190 *Official Year Book of the Commonwealth of Australia*, no. 46, 1960, Commonwealth Bureau of Census and Statistics, Canberra, p 951.
191 *Year Book*, no. 23, 1930, p 453.
192 *Year Book*, no. 46, 1960, p 951.
193 The Duke of Northumberland Papers, p 182. The descendants of the South African Cape horses were found to be smaller though hardier than the Indian stock which, being more 'hotly' bred with Arab blood, required better feed.
194 Alexander Harris, *Settlers and Convicts* (facsimile edition), Melbourne University Press, Melbourne, 1964, p 25. See also Nat Gould, *On and Off the Turf in Australia* (facsimile edition), Libra Books, Canberra, 1973, pp 104–5.
195 Curr, p 143.
196 ibid., p 144.
197 ibid., p 145.
198 ibid., pp 146–7.
199 ibid., pp 147–8.
200 SAPP, no. 67A, 1869, p 1.
201 Ron Iddon, 'The Australian stockhorse' in *The Stockman*, ed N. Mallon, Lansdowne Press, Sydney, 1988, p 207. See also pp 208–9.
202 ibid., p 209, and Curr, p 145.
203 Iddon, pp 209–10.
204 ibid., p 209.
205 A B Paterson, *Singer of the Bush*, Lansdowne Press, Sydney, 1985, p 61.
206 Curr, pp 25–32.
207 ibid., p 146.
208 Iddon, p 206. See also p 189.
209 SAPP, no. 67A, 1869, p 2.
210 George Fletcher Moore, 'Excursion to a river to the northward: From the journal of George Fletcher Moore,' Battye Library *Exploration Diaries*, vol 2, p 404.

211 John McDouall Stuart, *Explorations in Australia: The Journals of John McDouall Stuart*, 2nd ed., Saunders, Otley and Co, London, 1865, pp 177–8.
212 Curr, p 98.
213 This factor is closely examined and discussed in McLaren and Cooper, 'The saddle-horse,' *Journal of Australian Studies*, no. 49, 1966, pp 21–32.
214 Bruce Simpson, Battye Library accession no. OH 2730/10, Q12.
215 According to Edward Palmer, *Early Days in North Queensland* (facsimile edition), Angus and Robertson, Sydney, 1983, pp 261–3:
> 'Norman,' a large bay horse, bred on [Palmer's station] Conobie about 1870, was broken in three or four years after, and worked on till twenty-four or twenty-five years old as a stock horse, and then nearly as good and safe to ride as ever. A surer, better stock horse was never ridden, and always ridden by the writer.
216 Curr, pp 174–5. See also pp 249–51.
217 Paterson, *Song of the Pen*, pp 347, 350. See also p 277.
218 ibid., p 286; Curr, pp 89, 173–4.
219 SAPP, no. 25, 1874, p 4.
220 H M Barker, *Droving Days*, Hesperian Press, Perth, 1994, p 41.
221 Curr, p 118. See also pp 175–6, where Curr quotes Captain Shakespeare, who wrote in *Wild Sport in India* that Australian horses were 'the most difficult to break that can be found.'
222 ibid., pp 43, 46, 129.
223 Barker, pp 79, 81.
224 Reg Wilson, personal correspondence, 20 December 1997, p 27.
225 Simpson, Q11.
226 Curr, pp 115, 124–5, 129. See also Barker, pp 83–4, regarding his belief that Arabs bucked less than thoroughbreds while being broken in.
227 Coralie Gordon, 'The Australian Arabian performance horse' in *The Arabian Horse in Australia*, The Arabian Horse Society of Australia, 1983, p 14. Arabian and Anglo-Arab horses have taken a quite disproportionate share of honours in Australian endurance rides since the 1960s.
228 Curr, p 109.
229 Iddon, p 192.
230 Barker, p 13.
231 SAPP, no. 25, 1874, pp 4–5.
232 GRR, SAPP, no. 45, 1905, p 2.
233 SAPP, no. 67A, 1869, p 1.
234 Curr, p 141.
235 Paterson, *Song of the Pen*, p 346. See also Curr, p 110.
236 SAPP, no. 25, 1874, pp 3–4.
237 Barker, pp 84–6. Curr also admits (p 124) that Arabs are not as fast as thoroughbreds. Paterson is more explicit (*Song of the Pen*, p 277), noting that by the early 1900s 'no pure Arab, or other Eastern horse, would have any kind of chance of beating an English thoroughbred over any distance from

five furlongs to fifty miles.'

238 Curr admitted the pure Arab in India rarely exceeded fourteen hands high, and that the breed overall required 'two inches in height to be the perfection of horseflesh.'

239 Ion Idriess, *The Desert Column*, Angus and Robertson, Sydney, 1951, pp 217–26.

240 Ledger, Q65. See also Simpson, Q6.

241 Barker, p 83.

242 Jo Mora, *Trail Dust and Saddle Leather*, University of Nebraska Press, Lincoln, 1987, p 207.

243 Thoroughbreds have only six colours: bay, brown, black, chestnut, grey and white. These colours are fixed and broken colours cannot occur.

244 Mora, p 240.

245 The Morgan breed originated from one horse owned by Justin Morgan in Vermont at the end of the eighteenth century. This stallion stood only about fourteen hands high, but was possessed of a remarkable turn of speed as well as phenomenal strength and endurance.

246 Mora, pp 220, 242.

247 ibid., p 80.

248 During the BTEC eradication scheme all station cattle were mustered and systematically tested for bovine tuberculosis. Those testing positive were destroyed and at the same time potential carriers such as wild cattle, buffalo and horses were shot to prevent re-infection.

249 Reg Wilson, Battye Library accession no. OH 2730/12, Q43.

250 Glen Edgar, Battye Library accession no. OH 2730/4, Q79. This may be something of an exaggeration: rather than take the risk of souring them, new bronco horses were generally eased into work.

251 *Northern Territory Times*, 22 November 1884.

252 Barker, p 83.

253 Jack-donkeys are entire male donkeys.

254 Marcel Aurousseau, *The Letters of F W Ludwig Leichhardt*, vol 3, Cambridge University Press, London, 1968, p 919.

255 George Bates, 'An autobiography of George Bates,' unpublished memoirs, p 17.

256 Details supplied by the Royal Australian Artillery Historical Society, Inc (WA Branch).

257 Edgar, Q86 and Q87.

258 Bates, p 15.

259 Wilson, private correspondence, 20 December 1997, p 15.

Horsemanship

260 Simpson argues that bush horsemen saw husbandry as separate from horsemanship. Nevertheless, because the term horsemanship is derived from horseman (one who is proficient at the physical, psychological and physiological management of his horses), it will be used with this broader connotation.

261 W Thacker, Staff Veterinary Surgeon and remount agent of Calcutta, dismissed this notion, stating:

> In England sixty miles a day is no uncommon distance to ride or drive one horse, and eighty have not unfrequently been done by me in the hunting seasons. (SAPP, no. 25, 1874, p 3.)

This is not a reasonable comparison, for these horses were grain fed and were not asked to reproduce such work several days in succession.

262 Henry Handel Richardson, *The Fortunes of Richard Mahony: The Way Home*, Penguin Books Australia, Melbourne, 1976, p 16.

263 Nat Gould, *On and Off the Turf in Australia* (facsimile edition), Libra Books, Canberra, 1973, p 106.

264 W P Auld, *Recollections of McDouall Stuart*, Sullivans Cove, Adelaide, 1944, p 26.

265 Augustus Charles Gregory and Francis Thomas Gregory, *Journals of Australian Explorations* (facsimile edition), Hesperian Press, Perth, 1981, p 163.

266 A B Paterson, *Singer of the Bush*, Lansdowne Press, Sydney, 1985, pp 578–84. See also p 532. By contrast, Elyne Mitchell (*Light Horse*, Sun Books, Melbourne, 1987, p 67) quotes Gullet as saying that by 1917 the Australian light-horsemen had become 'superb horsemasters.'

267 A B Paterson, *Song of the Pen*, Lansdowne Press, Sydney, 1985, p 285.

268 Marcel Aurousseau, *The Letters of F W Ludwig Leichhardt*, vol 3, Cambridge University Press, London, 1968, p 982.

269 Jo Mora, *Trail Dust and Saddle Leather* University of Nebraska Press, Lincoln, 1987, p 124.

270 Rolf Boldrewood, *Robbery Under Arms*, ed Alan Brissenden, University of Queensland Press, Brisbane, 1979, p 71.

271 John Oxley, *Journals of Two Expeditions into the Interior of New South Wales* (Australiana Facsimile Editions no. 6), Libraries Board of South Australia, Adelaide, 1964, p 269. See also p 307.

272 Pearson, Huidekoper, Michener, Harbaugh, Law, Trumbower, Liautard, Holcombe, Adams, and Mohler, *Special Report of Diseases of the Horse*, United States Department of Agriculture, Bureau of Animal Industry, Washington, DC, 1942, pp 471–5.

273 Macquarie to Bathurst, 28 April 1814, Enclosure no. 4, *HRA*, S1 V8, pp 169–70. Presumably some notice was taken of this problem, for no later references can be found to pack-saddles not being correctly lined with serge.

274 Augustus Gregory, ms, Mitchell Library accession no. Q437, item 1.

275 Edward John Eyre, *Journals of Expeditions of Discovery into Central Australia*, vol

1, T and W Boone, London, 1845, p 32.
276 Reg Baker, Battye Library accession no. OH-2730/1, Q37. See also Bruce Simpson, Battye Library accession no. OH 2730/10, Q17.
277 Baker, Q37.
278 Oxley, p 247.
279 ibid., p 249.
280 ibid., p 256.
281 Gregory, ms, item 7, p 13.
282 See, for example, the Invoice of Provisions, *HRA*, S1V3, p 514. See also King to Portland, 10 March 1801, Enclosure no. 3, *HRA*, S1V3, p 72.
283 William Phillips, ms, Mitchell Library accession no. C 165, p 49.
284 Charles Sturt, *Journal of the Central Australian Expedition 1844–1845* (facsimile edition of Charles Sturt's *Expedition into Central Australia*, vol 2, 1849, ed Jill Waterhouse), Caliban Books, London, 1984, p 86.
285 Sturt, p 90
286 Gregory, *Journals*, p 72.
287 ibid., p 71.
288 ibid., p 75.
289 ibid., p 162.
290 Gregory, ms, item 1.
291 Compression and expansion of the rubbery frog also assists in the circulation of blood through the poorly vascularised lower leg.
292 It was also effective against horses that kicked at ropes, although the rope had to be laid in readiness on the ground and the horse walked over it into position. The ropes were then quickly pulled tight and the horse thrown before it had time to lash out.
293 Barker, pp 102–3.
294 ibid., p 103.
295 Lachlan Macquarie, *Journals of his Tours in New South Wales and Van Diemen's Land, 1810–1822* (facsimile edition), Library of Australian History, Sydney, 1979, p 145.
296 J A Simpson and E S C Weiner (eds), *The Oxford English Dictionary*, 2nd ed., vol 16, Clarendon Press, Oxford, 1989, p 102.
297 Alexander Harris, *Settlers and Convicts* (facsimile edition), Melbourne University Press, Melbourne, 1964, p 25.
298 Gregory, ms, item 1.
299 Ernest Giles, *Australia Twice Traversed: The Romance of Exploration*, 2 vols (Australiana Facsimiles Editions no. 13), Libraries Board of South Australia, Adelaide, 1964, vol 1, pp 288–9.
300 John McDouall Stuart, *Explorations in Australia: The Journals of John McDouall Stuart*, 2nd ed., Saunders, Otley and Co, London, 1865, p 47.
301 Eyre, vol 1, p 113.
302 GRR, SAPP, no. 54, 1885, p 2.
303 Pat Underwood, Battye Library accession no. OH 2730/11, Q59.

304 ibid., Q55.
305 Dave Ledger, Battye Library accession no. OH 2730/7, Q57.
306 Horses with leg injuries have traditionally been treated with mild to severely caustic mercuric compounds that blister and even burn the skin. The rationale behind this treatment is that the increased blood circulation resulting from the blister assists in healing the original injury.
307 Alexander Forrest, *North West Exploration*, Richard Pether, Government Printer, Perth, 1880, p 21.
308 Milton Moore (ed.), *Australian Grasslands*, Australian National University Press, Canberra, 1972, p 389.
309 The Messrs Jardine, *Narrative of the Overland Expedition of the Messrs Jardine, from Rockhampton to Cape York, Northern Queensland*, ed F J Byerley, J W Buxton, Brisbane, 1867, pp 20, 46.
310 Alfred Hillman, 'Mr Hillman's journal,' Battye Library *Exploration Diaries*, vol 4, p 350.

Saddlery and Harness

311 William Charles Wentworth, *Statistical, Historical, and Political Description of the Colony of New South Wales and its Dependent Settlements in Van Diemen's Land* (facsimile edition), Griffin Press, Adelaide, 1978, p 111.
312 George Bates, 'An autobiography of George Bates,' unpublished memoirs, p 25.
313 ibid., p 22.
314 *The Qld (Wise's) Official Directory 1896–97*, H Wise and Co, Brisbane, 1896, pp 1085–6, 1157.
315 According to ninety-three-year-old saddler Bob Cameron of Morley, Perth, he and his father worked as travelling saddlers in the Kimberleys during the 1930s. Each year they drove to Derby and surrounding stations, then used packhorses to reach those properties not connected by road.
316 Jim Hill, Battye Library accession no. OH 2730/6, Q4.
317 Don Bates, Battye Library accession no. OH 2730/2, Q5, Q6 and Q7.
318 Edward M Curr, *Recollections of Squatting in Victoria*, George Robertson, Melbourne, 1883, p 117.
319 A B Paterson, *Singer of the Bush*, Lansdowne Press, Sydney, 1985, pp 583–4.
320 George Bates, p 45.
321 Using a saddle with a broken tree is not recommended, for it can cut a horse's back or withers badly.
322 Joseph Furphy, *Such is Life*, Angus and Robertson, Sydney, 1975, pp 345–6.
323 A B Paterson, *Song of the Pen*, Lansdowne Press, Sydney, 1985, p 661.
324 Reverend William Haygarth MA, *Recollections of Bush Life in Australia, During a Residence of Eight Years in the Interior*, John Murray, London, 1864, p 78.
325 A crupper is a leather strap connected to the rear of a saddle, which passes in

a loop underneath the horse's tail. When adjusted correctly, it prevents the saddle slipping forward over the wither.
326 The top section of the front of a saddle.
327 Edward M Curr, *Pure Saddle Horses, and How to Breed Them in Australia*, Wilson and MacKinnon, Melbourne, 1863, p 178.
328 Eventually, because of problems of affixing them to the side panels, they reached their upper limit. Not until the advent of fibreglass trees, on which pads could be moulded at any position and merely covered with leather, was this limitation overcome.
329 George Bates, p 22.
330 ibid., p 21.
331 Jo Mora, *Trail Dust and Saddle Leather*, University of Nebraska Press, Lincoln, 1987, p 98.
332 By the late 1940s, however, the original slick format was regaining popularity. See Mora, pp 92–6, for further details.
333 ibid., p 106. Horsebreakers who had lost their nerve to the extent that they relied on roughened saddles would have been riding neither confidently nor well. Most of the benefit gained would have been psychological but, so long as the rider relaxed and concentrated on the task rather than on the likelihood of falling, he would have been riding far more easily and effectively.
334 Bates, p 44.
335 Hill, Q1 and Q2.
336 Hill was not the first to use felt panels. Frank Dean recalls drover, roughrider and amateur jockey Ted Young of Kynuna riding in a Schneider saddle with felt panels in 1940.
337 Hill, Q1 and Q2. Felt panels did not completely solve the problem of sore backs, for the felt had to be regularly scrubbed to remove dried sweat and hair. Furthermore, according to Simpson they often had to be replaced after two or three seasons of hard work.
338 ibid., Q13.
339 R M Williams, 'Australian bush crafts,' in Nola Mallon (ed.), *The Stockman*, Lansdowne, Sydney, 1988, p 177.
340 Jack Gardiner, *Stockman's Hall of Fame*, September 1993, p 6.
341 W H L Ranken, *The Dominion of Australia: An Account of its Foundations*, Chapman and Hall, London, 1874, p 124.
342 Hill, Q10.
343 Mora, p 117.
344 Thomas Bannister, 'Extracts from a journal kept by Thomas Bannister in charge of a party from Swan River to King George's Sound,' Battye Library *Exploration Diaries*, vol 1, p 231.
345 John Septimus Roe, 'Report by J S Roe Esq: Surveyor General, of his expedition to explore the interior country south eastward from York, between September 1848 and February 1849,' Battye Library *Exploration Diaries*, vol 4, p 151.
346 W B Lord and T Baines, *Shifts and Expedients*, pp 30–1, cited in Wendy Birman,

Gregory of Rainworth, A Man in his Time, University of Western Australia Press, Perth, 1979, pp 119–20.
347 Augustus Charles Gregory and Francis Thomas Gregory, *Journals of Australian Explorations* (facsimile edition), Hesperian Press, Perth, 1981, p 120.
348 ibid., p 198.
349 William Landsborough, *Journal of Landsborough's Expedition from Carpentaria, in Search of Burke and Wills: With a Map Showing his Route* (facsimile edition), State Libraries Board of South Australia, Adelaide, 1971, p 23.
350 Alexander Forrest, *North West Exploration: Journal of an Expedition from DeGrey to Port Darwin*, Richard Pether, Government Printer, Perth, 1880.
351 See Mora, pp 91–7, for a comprehensive outline of US saddlery.
352 ibid., pp 100–2.
353 ibid., p 104.
354 ibid., p 75. Mora deals with US bits comprehensively on pp 67–85.
355 This extravagance can perhaps be traced to the Mexican origins of much US saddlery.
356 Mora, p 101.
357 ibid., p 107.

Horsebreaking

358 Jo Mora, *Trail Dust and Saddle Leather*, University of Nebraska Press, Lincoln, 1987, pp 71–2.
359 ibid., p 81.
360 Physical immaturity was often due to the inadequate nature of much of the available pasturage. The late breaking-in was also based on the horses' very heavy workload. As former Territory breaker Reg Baker recalls:
> Out in Western Queensland and the Northern Territory horses were usually never touched for breaking in until they were the age of five. The reason was that they had to cover enormous distances in the day's muster, say 70 miles a day. If you tried to do that with a three year old you'd break him down in no time. They just couldn't stand the work. (Battye Library accession no. OH 2730/1, Q23)

361 *Northern Territory Times*, 27 July 1918.
362 Thomas Cockburn-Campbell, *Land of Lots of Time*, Fremantle Arts Centre Press, Fremantle, 1985, p 138. See also Charlie Schultz (*Beyond the Big Run*, ed Darrell Lewis, University of Queensland Press, Brisbane, 1996, p 13) for his description of how a patronising English veterinarian was struck to the ground by a horse in India in 1925.
363 Pat Underwood, Battye Library accession no. OH 2730/11, Q25 and Q26.
364 Baker, Q24 and Q25.
365 Campbell, p 91. By contrast, US cow ponies were carefully educated before being sent to the stock camps. Differing requirements appear to underlie this

approach, for the average US cowboy was almost certainly as good a rider as the Australian stockman. Whereas Australian horses had to be able to stop and turn quickly, and to gallop alongside cattle that were to be thrown, US horses had also to be practised at roping procedures — they had to slide to a stop when a beast was roped and stand stationary, keeping the rope taut as the rider dismounted and ran towards the struggling animal.

366 Mrs Aeneas Gunn, *We of the Never-Never*, Hutchinson, London, 1962, pp 83–8. See also Baker, Q20, Q32 and Q36.
367 Henry Kingsley, *The Recollections of Geoffry Hamlyn*, ed J S D Mellick, University of Queensland Press, Brisbane, 1982, p 146.
368 SAPP, no. 25, 1874, p 5.
369 Underwood, Q58.
370 See Campbell, p 11, for a heart-rending description of this technique.
371 Schultz, pp 23–4.
372 Campbell, p 121.
373 ibid., p 110.
374 Bruce Simpson, *Packhorse Drover*, ABC Books, Sydney, 1995, p 133.
375 Mouthing is most definitely not an automatic process. If a horse decides it will not cooperate and becomes sulky and stubborn, almost no degree of force or brutality can make it. The breaker has to use skill and ingenuity, and if the animal is not cooperating satisfactorily with one approach another must be tried. See Baker, Q22 and Q25. See also Underwood, Q25, Q26 and Q27.
376 H M Barker, *Droving Days*, Hesperian Press, Perth, 1994, p 95.
377 ibid., p 96.
378 Schultz, p 24.
379 Campbell, pp 92–3.
380 Barker, p 94.
381 Better handled horses were not so dangerous, and Simpson, for one, rasped his droving plant's teeth at the beginning of each season.
382 According to Mora, p 227, twitches are given the same name and used for the same purpose in the US.
383 A B Paterson, *Song of the Pen*, Lansdowne Press, Sydney, 1985, p 657.
384 ibid., p 661. This is a wonderfully descriptive passage of writing.
385 ibid., p 662.
386 Joseph Furphy, *Such is Life*, Angus and Robertson, Sydney, 1975, p 63.
387 Mora, p 67.
388 Schultz, pp 113–4. See also Campbell, pp 123–4, for details of how he and an associate held a horse down by the ears and bridle as it was being mounted.
389 Mora, pp 118–9.
390 Furphy, p 350.
391 Ironically, the first three commentators are contradicting one another.
392 This is something of an exaggeration. As Simpson points out, 'the horse has not been foaled that can buck badly for sixty seconds.'
393 Furphy, pp 351–2.

394 Simpson, p 28.
395 W T Pike, ed., *Bush Tales, by Old Travellers and Pioneers*, cited in *S T Gill's Rural Australia*, ed R. Raftopoulos, Oz Publishing Co., Brisbane, 1987, p 32.
396 Tom Cole, *Hell West and Crooked*, Collins Publishers Australia, Sydney, 1988, p 15. See also p 51, where Cole admitted on the occasion he had to ride a horse whose 'milk of equine kindness ran very thinly in his veins' that 'frankly I had the shit scared out of me.'
397 Campbell, p 124. See also p 186.
398 Mora, p 96.
399 ibid., p 100.
400 A stout, short-legged riding horse.
401 Furphy, p 173.
402 Barker, pp 99–100.
403 Peter Ross, Battye Library accession no. OH 2730/9, Q40.
404 Nehemiah Bartley, *Australian Pioneers and Reminiscences 1849–1894* (facsimile edition), John Ferguson, Sydney, 1978, p 195.

Draught Animals

405 Edward Palmer, *Early Days in North Queenland* (facsimile edition), Angus and Robertson, Sydney, 1983, p 192. See also Nehemiah Bartley, *Australian Pioneers and Reminiscences 1849– 1894* (facsimile edition), John Ferguson, Sydney, 1978, p 199.
406 H M Barker, *Camels and the Outback*, Pitman, Melbourne, 1965, p 189.
407 *Northern Territory Times*, 13 September 1919.
408 Barker, p 77.
409 Hermann Beckler, *A Journey to Cooper's Creek*, ed Stephen Jeffries, Melbourne University Press, Melbourne, 1993, pp 43–4.
410 Barker, pp 170–1.
411 ibid., p 167.
412 ibid., p 170.
413 Beckler, p 132. See also Barker, p 73, where he recounts having to walk fifty miles to find and return with his camels.
414 Barker, p 13.
415 Chains were always given close attention, for if a link wore thin and snapped while the team was pulling, the animal could cannon forward and crash into the one in front. These best and hardest pullers could subsequently be so frightened they would never pull truly again. Thus worn links were cut out by the blacksmith and replacements welded in.
416 ibid., pp 206–7.
417 Beckler, p 142.
418 Gelded camels had the further advantage that they did not smell so badly as bulls.

419 Barker, p 20.
420 Arthur Cannon, *Bullocks, Bullockies and Other Blokes*, Hill of Content, Melbourne, 1983, p 61.
421 White animals were, however, more prone to skin cancer and photosensitisation.
422 Cannon, p 191.
423 Joseph Furphy, *Such is Life*, Angus and Robertson, Sydney, 1975, p 268
424 'Leisure,' *Sunday Times*, Perth, 9 February 1997, p 11.
425 ibid.
426 Bullwhips were rarely used in Australia. They were commonplace in the south of North America, Mexico and South America.
427 Thomas Cockburn-Campbell, *Land of Lots of Time*, Fremantle Arts Centre Press, Fremantle, 1985, p 116.
428 Barker, p 18.
429 Edward John Eyre, *Journals of Expeditions of Discovery into Central Australia*, 2 vols, T and W Boone, London, 1845, vol 1, p 37.
430 Barker, p 70.
431 Mrs Aeneas Gunn, *We of the Never-Never*, Hutchinson, London, 1962, pp 131–2.
432 ibid., p 43.
433 Edmond Marin La Meslee, *The New Australia*, Heinemann, London and Melbourne, 1973, pp 38–9.
434 ibid., pp 127–30.
435 Barker, p 200.
436 Palmer, p 191.
437 Barker, p 8.
438 ibid., p 81.
439 ibid., p 85.
440 Barker, p 198.
441 ibid., pp 193, 199.

Safety and the Use of Firearms

442 Henry Reynolds, *The Other Side of the Frontier*, Penguin Books, Melbourne, 1990, pp 121–2.
443 George Serocold, *Sydney Morning Herald*, 30 November 1857, cited in Gordon Reid, *A Nest of Hornets*, Oxford University Press, Melbourne, 1982, p 73.
444 Reynolds, p 61.
445 Report of Assistant-Protector E S Parker, 20 June 1839, cited in Reynolds, p 113.
446 It can be argued that Australian settlers were better placed than their US counterparts. In the South they had to face both dispossessed Mexicans and the armed and formidably efficient indigenous Indians who, like the Australian Aborigines, had no compunction about indiscriminate murder.

447 Initially, under the highly effective 'keeping out' policy, Aborigines were driven well away from homesteads into broken, inaccessible country which was of little value to grazing stock. When they were considered sufficiently trustworthy, they were allowed to 'come in'. The practice is described in greater detail later in this chapter.

448 Ion Idriess, *Over the Range*, Angus and Robertson, Sydney, 1942, p 243. Ann McGrath (*Born in the Cattle*, Allen and Unwin, Sydney, 1987, p 11) outlines how Alec McDonald was murdered in the Kimberleys in 1918 because 'he had plenty of tobacco.'

449 Alfred Searcy, *In Australian Tropics* (facsimile edition), Hesperian Press, Perth, 1984, p 191.

450 Reid argues (pp 57, 63) that the Fraser women were raped before being murdered in the Hornet Bank massacre in revenge for the abuse of Aboriginal women by the Fraser sons.

451 Footnote to Barrallier's journal, 10 December 1802, *HRN*, V5, p 815.

452 Reynolds (pp 77–8) addresses the Aboriginal notion of revenge and the white response. Reid (pp 188–9) deals succinctly with the incompatibility of these two systems of justice and how blacks and whites alike gradually abandoned their traditional measured responses.

453 Major T L Mitchell, *Journal of an Expedition into the Interior of Tropical Australia in Search of a Route from Sydney to the Gulf of Carpentaria*, Longman, Brown, Green and Longman, London, 1848, p 253. It should be noted that Mitchell also pointed to mitigating factors.

454 Reid, pp 53, 61–3.

455 Charlie Schultz, *Beyond the Big Run*, ed Darrell Lewis, University of Queensland Press, Brisbane, 1996, pp 44–5. By contrast, McGrath (pp 11–12) states that Ward was murdered for threatening to shoot an elderly Aboriginal man and his young wife if the husband would not agree to the woman cohabiting with Ward.

456 *Northern Territory Times*, 14 June 1895. See Idriess, pp 59–62, for details of how Kimberley Aborigines painstakingly fashioned glass spear-heads, using tools hand-made from fencing wire, and grindstones.

457 See McGrath, pp 18–23.

458 ibid., pp 5–6.

459 George Boxall, *The Story of the Australian Bushrangers* (facsimile edition), Penguin Books Australia, Melbourne, 1974, pp 27–9. McGrath notes, however, (pp 8–9) that, because Aborigines were not a united people fighting for a common cause, it is misleading to speak of guerilla warfare. The analogy is nevertheless a useful one.

460 Schultz, pp 1–2.

461 Idriess, p 36.

462 Searcy, pp 19–21.

463 Mrs Aeneas Gunn, *We of the Never-Never*, Hutchinson, London, 1962, p 72. Schultz records (pp 165, 206) that even in the early 1940s revolvers were still carried.

464 *Adelaide News*, 7 July 1929.
465 Percussion caps were sealed capsules, containing highly explosive potassium chloride, which were ignited when struck by a spring-loaded firing pin or hammer, thus setting off the powder charge in the barrel and expelling the ball or shot. Although their use marked a considerable advance on the flintlock technique, where misfires were common, even the sealed caps were deleteriously affected by high humidity.
466 Dr Ludwig Leichhardt, *Journal of an Overland Expedition in Australia, from Moreton Bay to Port Essington* (facsimile edition), Macarthur Press, Sydney, 1847, p 309.
467 Searcy, p 227.
468 Caley to Banks, 12 October 1800, *HRN*, vol 4, p 240.
469 Boxall, p 53,
470 A M Low, *Musket to Machine-Gun*, Hutchinson, London, 1942, p 22.
471 H M Barker, *Droving Days*, Hesperian Press, Perth, 1994, pp 31–2.
472 Attributed to James Twigg of Western Australia in 1891 and cited in Patrick O'Farrrell, *Letters from Irish-Australia 1825–1929*, ed B. Trainor, New South Wales University Press, Sydney, 1984, pp 86–7.
473 W H Tietkens, Mitchell Library accession no. 1352–2, pp 97, 107.
474 Jo Mora, *Trail Dust and Saddle Leather*, University of Nebraska Press, Lincoln, 1987, p 47.
475 *Northern Territory Times*, 19 August 1885.
476 See Reid, pp 3–4, 36–7, 40, for details on the establishment of the Native Police and the location of the initial four divisions.
477 *Port Denison Times*, 20 November 1869, cited in Noel Loos, *Invasion and Resistance*, Australian National University Press, Canberra, 1983, p 33. According to Reid (p 134), following the Cullen la Ringo massacre at Comet in 1861, Native Police Commandant Bligh claimed it was a 'well-known fact … that no murders were committed at any station where the Blacks were entirely kept out.' Accordingly, he directed that no Aborigines were permitted to remain on any station in the Comet district.
478 The Cullen la Ringo massacre occurred approximately ten miles south of Springsure, Queensland. Certainly there was little public sympathy, nor means of redress, for Aborigines. Their evidence was inadmissible in courts, and jurors, especially in rural areas, were reluctant to convict white settlers.
479 Loos, pp 49, 58–9. These properties were Robert Christison's Lammermoor and William Chatfield's Natal Downs. Chatfield claimed he could not run his property without the Aborigines' assistance.
480 ibid., p 102.
481 *Northern Territory Times*, 11 October 1901.
482 Reid, pp 87–91, 113, 181–3.
483 See Idriess, pp 36, 198–9, McGrath, p 107, Tom Cole, *Hell West and Crooked*, Collins Publishers Australia, Sydney, 1988, p 179, and A J Macgeorge, *Where Fortunes Lay*, Hesperian Press, Perth, 1993, p 128.
484 Schultz, p 55.

485 Idriess, pp 36–7.
486 ibid., p 198. According to historian Ann McGrath (p 107), 'an old trick to get dogs to attack blacks on sight [was that] … a dog was tied in a bag, belted around "a bit," [and] then an Aborigine would be made to release it.' McGrath, p 11, and Cole, p 179, also provide details of drovers and travellers being warned by their dogs who had scented Aborigines in close proximity.
487 Searcy, p 256.
488 ibid., p 257.

Homesteads, Yards and Fences

489 Thomas Cockburn-Campbell, *Land of Lots of Time*, Fremantle Arts Centre Press, Fremantle, 1985, p 135.
490 James Broadbent, *The Australian Colonial House*, Hordern House, Sydney, 1997, p 19.
491 Cockburn-Campbell, pp 149–50.
492 Mrs Aeneas Gunn, *We of the Never-Never*, Hutchinson, London, 1962, p 124.
493 Alfred Searcy, *In Australian Tropics* (facsimile edition), Hesperian Press, Perth, 1984, p 327. Northern orchardists gradually learnt that regular ploughing eventually kills white ants and prevents reinfestation.
494 ibid., p 323
495 ibid., p 11.
496 Charlie Schultz, *Beyond the Big Run*, ed Darrell Lewis, University of Queensland Press, Brisbane, 1996, p 186.
497 Reg Wilson, Battye Library accession no. OH 2730/12, Q13.
498 ibid., Q18.
499 Ray Summers, 'The Queensland style' in *The History and Design of the Australian House*, ed Elizabeth Dan, Oxford University Press, Melbourne, 1985, p 303.
500 Gunn, p 54.
501 ibid., pp 54–5.
502 Annette Henwood, Battye Library accession no. OH 2730/5, Q31.
503 ibid., Q28 and Q29.
504 Gunn, pp 124–5.
505 Henwood, Q44.
506 ibid., Q44.
507 Wilson, Q4 and Q10.
508 *The Stockowner's Guide,* Pastoralists' Review, Sydney, 1912, p 122.
509 R A Billington, *Westward Expansion*, 2nd ed., The Macmillan Company, New York, 1960, pp 691–2.

Rangeland Management

510 Augustus Charles Gregory and Francis Thomas Gregory, *Journals of Australian Explorations* (facsimile edition), Hesperian Press, Perth, 1981, pp 39–40.

511 *Draft National Strategy for Rangeland Management*, Department of the Environment, Sport and Territories, Commonwealth of Australia, 1996, p 9. This not only refers to flora, but also includes alterations to fauna populations and environment.

512 ibid., p 28.

513 Burning was also seen as beneficial in controlling ticks. According to Reg Wilson, Dorisvale station carried out this practice from the 1880s onwards.

514 Reg Wilson, Battye Library accession no. OH 2730/12, Q1 and Q2.

515 Annette Henwood, Battye Library accession no. OH 2730/5, Q1.

516 J H Kelly, *The Beef Cattle Industry in the Leichhardt–Gilbert Region of Queensland*, Bureau of Agricultural Economics, Canberra, 1959, p 33.

517 Kelly, pp 32–3, has reservations regarding this claim.

518 Henwood, Q8.

519 ibid., Q2 and Q4.

520 Wilson, Q1.

521 *Draft National Strategy*, p 8.

522 Kelly, p 33.

523 Nehemiah Bartley, *Australian Pioneers and Reminiscences 1849–1894* (facsimile edition), John Ferguson, Sydney, 1978, pp 275–6.

524 A B Paterson, *Singer of the Bush*, Lansdowne Press, Sydney, 1985, p 260.

525 Edward Palmer, *Early Days in North Queensland*, Angus and Robertson, Sydney, 1983, p 230.

526 The leader walked alone alongside the previous furrow to ensure the furrows were straight and constant.

527 Frank Dean, Battye Library accession no. OH 2730/3, Q7.

528 Simon Fraser, 'The true story of the beginning of the artesian water supply of Australia,' *Stockman's Hall of Fame*, June 1996, p 6.

529 Palmer, p 103.

530 ibid., pp 228–31.

531 For a more detailed outline see M A Habermehl, *Investigation of the Geology and Hydrology of the Great Artesian Basin, Report 234*, Bureau of Mineral Resources, Geology and Geophysics, NTU SC 551.490994 HABE.

532 Fraser, p 6.

533 Palmer, pp 229–30.

534 Dean, Q20.

535 Palmer, p 229.

536 *The Stockowner's Guide*, Pastoralists' Review, Sydney, 1912, p 76.

537 To overcome such losses, proposed remedies included baking the channel soil by applying heat, stabilising it by adding cement or asphalt, or rendering it

impermeable by incorporating sodium carbonate. See Habermehl, p 18.
538 Paterson, p 267.
539 Habermehl, p 9.
540 Felix Schmidt, 'Memorabilia of "Alroy Downs," NT,' unpublished paper, 1990, pp 1, 3.
541 Dean, Q5.

Equipment and Know-how

542 Darling to Hay, 22 December 1827, in an enclosure from Hely to McLeay, *HRA* S1 V13, pp 663–4.
543 A B Paterson, *Song of the Pen*, Lansdowne Press, Sydney, 1985, p 660.
544 Jo Mora, *Trail Dust and Saddle Leather*, University of Nebraska Press, Lincoln, 1987, p 44. An illustrated outline of the development of the US cowboy boot appears on p 45. See also Jim Hoy, *Cowboys and Kansas*, University of Okalahoma Press, Norman, 1995, pp 193–8, 205–6.
545 By contrast, US ropers using a roping horn at times had to hold and manipulate a rope that had a grown cow or bullock on the other end.
546 *The Stockowner's Guide,* Pastoralists' Review, Sydney, 1912, p 225.
547 ibid., p 225.
548 A spreader bar kept the three strands apart in the middle and prevented them from becoming tangled.
549 Brian Nemath, Battye Library accession no. OH 2730/8, Q23.
550 ibid., Q17.
551 ibid., Q24.
552 *Stockman's Hall of Fame*, June 1984, p 11.
553 ibid., p 11.
554 Alfred Searcy, *In Australian Tropics* (facsimile edition), Hesperian Press, Perth, 1984, p 130.
555 Mora, p 106.
556 Thomas Cockburn-Campbell, *Land of Lots of Time*, Fremantle Arts Centre Press, Fremantle, 1985, p 155.
557 Cited in J H L Cumpston, *Charles Sturt*, Georgian House, Melbourne, 1951, Appendix B, pp 170–4.
558 Glen Edgar, Battye Library accession no. OH 2730/4, Q61.
559 Arthur Cannon, *Bullocks, Bullockies and Other Blokes*, Hill of Content, Melbourne, 1983, p 160.
560 Searcy, p 209.

Epilogue

561 Geoffrey Bolton, *Hindsight*, vol 6, no. 1, June 1966. See also Sue King for her comments on students' and teachers' perceptions.
562 *Year Book of Australia*, no. 77, Australian Bureau of Statistics, Canberra, 1994, p 94.
563 Donald Horne, 'Look to museums to find yourself,' *The Weekend Australian*, 5–6 January 1992.

Conversion Table

Imperial to metric:

> one inch = 2.54 centimetres
> one foot: 12 inches = 30.5 centimetres
> one yard: 3 feet = 0.91 metre
> one mile: 1760 yards = 1.61 kilometres
> one acre = 0.41 hectare
> one quart = 1.14 litres
> one gallon: 4 quarts = 4.55 litres
> one pound = 0.45 kilogram
> one ton = 1.02 tonnes

Monetary:

> one penny = 0.83 cents
> one shilling: 12 pence = 10 cents
> one pound (£1): 20 shillings = 2 dollars

In noting the prices and costs quoted in this book, it should be borne in mind that inflation has progressed at a considerable rate since Australia was colonised. For example, over the past century the currency has diminished in value forty- to fifty-fold, meaning that £10 one hundred years ago is roughly equivalent to $1000 in today's money.

Glossary

Antbed — Antbed floors are constructed of pulverised white-ant mound, which is a mixture of clay and partially digested dried wood. The dry pulverised mixture is placed in position, levelled, wetted and tamped firm. When dry, it sets hard and forms a reasonably dust-free floor.

Appaloosa — An American breed of horse descended from the original Spanish stock and subsequently adopted by the Nez Perce Indians. They exhibit six coat colour combinations and white or dark spots are common, particularly over the hindquarters.

Bang-tail muster — Where an accurate count was required, as cattle were mustered they often had the hair at the base of their tails cut off square, half-way down. By this technique musterers could quickly see if a small mob had already been counted.

Barkly Tableland — A flat plateau region of superb grasslands in the north-east of the Territory, immediately inland from the coastal plain.

Blacksoil — A particularly fertile black soil with a high clay content which cracks as it dries. The cracks allow dry plant material to lodge below ground level and thus these soils are self-mulching.

Blister — Caustic substances such as mercuric compounds or petrol, which are applied superficially to tendon or ligament injuries of horses. The resultant severe blistering is reputed to increase blood circulation to the affected area and thus accelerate healing of the injury.

Branding crush — A mechanical device installed at the end of raceways, which locks onto a beast's head and immobilises it, thus allowing stockmen to carry out marking operations in safety.
Breakaway country — Rough, broken and often rocky country, which is frequently found at the junction of coastal plains and plateaus.

Breaking tackle — composite term for breaking-in equipment, which includes the breaking-in roller, breastplate and crupper.

Breasting or breastplate — a leather strap which is attached to either side of a saddle girth or roller strap, and runs forward around the horse's chest. Its function is to prevent the saddle or roller from slipping backwards.

Breeching or britching — a leather strap which is attached to the back of a pack saddle and runs back around the horse's hindquarters. Its function is to prevent the pack saddle from slipping forwards.

Brigalow — A medium to tall tree of the Acacia species, found in southern Queensland.

Broken country/terrain — rough, rocky and often hilly country (see breakaway country)

Bronco — A US term for a wild or poorly broken horse.

Bronco panel — A solid rail construction against which roped beasts are dragged by a bronco horse, for ear marking, castrating and branding.

Bronco rope — A rope used in the broncoing process, with which a lassoed beast is dragged to the bronco panel.

Broncoing — A Mexican practice, whereby selected beasts are lassoed in a mob and dragged forward by a bronco horse to the bronco panel for marking. This process was adopted in the Channel country, the Territory and the Kimberleys from the early 1900s onwards, and used until the mid to late-1960s

Brumby — An Australian term for a feral horse.

Bucking roll — In this context, a mackinaw was strapped on to the centre front of the saddle, in an attempt to change the configuration of the saddle and thus assist the rider with a horse which he believed would buck.

Bulldogger — A person who throws a galloping beast by taking it by the horns and twisting its head sidewards, to the extent that the animal falls.

Bull — An entire male of the bovine species.

Bullock — A castrated male of the bovine species.

Cantle — the hind part of a saddle.

Cape — Cape York Peninsula, or Capetown, South Africa.

Channel country — A region of relatively flat country criss-crossed with shallow watercourses in the mid-western region of Queensland, which is inundated when the Coopers Creek floods. The floodwaters cover huge areas of fertile soil and the resultant feed is of superb quality. This region is renowned as fattening country.

Chaps — Leather or sheepskin coverings worn by US cowboys over their trousers to provide protection from thorns, sweat and cold.

Cheekstrap — That section of a bridle which runs along the cheeks and connects the bit to the headpiece.

Close selection — The Queensland policy whereby large stations were broken down into a number of smaller properties, to enable more homesteads to be established.

Clydesdale — A medium to heavy British breed of draught horse.

Coacher — Coachers are quite, well-handled cattle used by musterers to decoy wild cattle, and to help control them.

Cowkick — A kicking motion whereby a horse or cow brings its hindfoot forward and then kicks backward and outward.

Crush — See Branding crush.

Curb bit — Often used on horses which pull hard, they are comprised of a standard bit and a small chain which hooks underneath the horse's lower jaw. Pulling on the reins causes this chain to press firmly against the horse's sensitive tissue and the resultant pain makes it more willing to comply with the rider's wishes.

Cut out — The process where a rider enters a mustered mob on a specially trained horse and selects a particular beast. This animal is then gradually shouldered and forced out of the mob and driven to a separate mob of selected animals (ie the 'cut.')

Delver — Bore drain delvers have a grading action and scour bore drains clean of weeds and silt.

Depasture — to graze stock on a particular area.

Draft — The process of sorting stock into differing groups (ie to draft cattle). Alternatively, a group of stock selected from the main mob according to specific criteria.(ie. a draft of fat bullocks.)

Dropper — Unlike fence posts which are fixed in the ground, dropper posts are merely suspended from the wire. Generally there are two or three droppers between the fixed posts. Their function is to keep the wires the correct distance apart, yet save the work and expense of additional fixed posts.

Dry-daying — A droving practice adopted when water was short, in which the stock received a drink only on alternate days.

Dun — A term used to describe dull greyish-brown or sandy-brown horses.

Farrier — A person who shoes horses, as distinct from a blacksmith who carries out general steelwork

Flank girth or balance girth — A second girth mounted eight to ten inches to the rear of the main girth. It is used to prevent the back of the saddle rising as a horse bucks, and thus the saddle slipping forward onto the horse's neck.

Forcing yards — Are smaller than the yards which initially receive the stock. Consequently the stock have less room to mill around in and can be forced more readily into the race, dip or drafting pens.

Frog — A triangular, rubber-like structure on the bottom of a horse's hoof.

Front-legging — A method of throwing cattle by the front leg, rather than by the tail or horns.

Galling — Raw wounds caused by ill-fitting saddlery, overworking an animal before skin in contact with the saddlery has become toughened through work, or excessive sweating arising from working in oppressively hot or humid conditions.

Girth — A broad leather strap attached to saddlery, which passes below an animal's rib-cage and which, when tightened, holds the saddlery in position. Alternatively, the area on a horse or pack-bullock immediately behind the front legs where the girth (strap) is located.

Greenhide — Uncured hide which is used for a variety of purposes, including rope-making, saddlery and furniture.

Gulf — Generally refers to the coastal plain adjoining the Gulf of Carpentaria.

Gunny sack — A coarse sack, usually jute.

Hackamore — A bridle without a bit. The rider steers the horse through pressure applied to the nose by the reins.

Halter — Leather or rope harness which is fitted around a horse or bullock's head, and which provides control over the animal.

Hames — Two curved pieces of iron or wood, which form part of the collar of a draught horse.

Headstall — see halter.

Hobble or hopple — A short length of chain with leather straps attached at each end, which are buckled to a horse or bullock's pasterns (ie lower leg). As a result the animal can only take short steps and cannot wander too far when turned out to graze overnight.

Honest (working animals) — Those animals which take their share of the load.

Horse plant — A team of working horses.

Kimberleys — Western Australian term for the Kimberley region.

Lariat — American term for a rope used to lasso stock.

Lunge or longe — The practice of running a horse around in a circle at the end of a rope, before it is ridden, in order to remove any bucks before the rider mounts it.

Maverick — A US term for an unbranded calf or yearling. Also used to describe a horse, often in the sense that it is a fiery animal.

Mesquite — A North American leguminous tree, which has exceedingly long, sharp thorns. It has been released in the north of Australia, where it has become a serious pest for stock and horsemen.

Mouldboard — A form of plough which, unlike Australian disc ploughs, turns the soil over in a continuous strip.

Near side/off side — The near side is an animal's left hand side, while the off side is its right hand side. Traditionally, stock are handled from the near side.

Neatsfoot oil — 'Neat' refers to bovines. Thus neat's-foot oil (the proprietary name of which is Neatsfoot Oil) presumably refers to oil rendered from cattle hooves.

Overshoot — A wall constructed of naturally occurring flagstones, built in watercourses downstream from permanent waterholes, which dams water back and increases the amount of water held.

Oxbow — A crude form of stirrup of American origin.

Palomino — A horse the colour of 'newly minted gold coin,' with pure white mane and tail.

Pastern — That bone in a horse's lower leg between the fetlock and hoof.

Percheron — Light to medium draught horses which originated in France. They do not have a feather on the legs, and are surprisingly supple and agile for their size.

Piebald — A horse exhibiting black and white colouring.

Pilbara — That region of Western Australia north of the Gascoynes and south of the Kimberleys. It embraces Marble Bar, Port Hedland, Roebourne and Wittenoom.

Pindan — Red, clay-rich soil prevalent in the Pilbara and Murchison regions.

Pise — Building material comprised of rammed earth, clay and possibly gravel.

Pit-sawing — When logs were cut by hand deep pits were dug underneath felled trees. One sawman then worked standing on the log, while the second (unfortunate) sawman worked below in the pit. Using a crosscut saw vertically, they cut the logs longitudinally into the desired thicknesses.

Plant — See horse-plant.

Pommel — The front top-centre part of a saddle.

Prickly-pear — A tall, prickly form of cactus brought to Australia at or shortly after settlement. Once this cactus reached northern New South Wales and Queensland it thrived in the warmer climate and spread unchecked, choking out thousands of square miles of pasture.

Purlin — A horizontal beam resting on the principal roof bearers, which supports the rafters.

Quarter-horse — A USA breed of horse specially bred for speed over short distances, docility and agility, which are renowned for their cattle working skills.

Rassle — To wrestle.

Reata — Lassoing rope.

Redhide — Cowhide tanned with alum.

Remount — Horses sent to India as mounts for the Army.

Roan — A horse whose predominant coat colour is thickly intermingled with a second colour. In Australia this is taken to be a chestnut with grey flecks.

Roller — A broad leather strap used for breaking in, which is fitted around the horse's girth.

Roping pole — A long thin pole with a small fork at one end, along which stockmen ran a rope, with a noose dangling from the fork. By this means they were able to place the noose closer to an animal's head, preparatory to lassoing it.

Rowel — A spiked revolving disc at the end of a spur, which, when applied, pricks the horse.

Rush — When cattle panic and run blindly as a herd. Known as a stampede in the USA.

Saddle flap — Large flaps of leather hanging down each side of a saddle, which protect the rider's legs from sweat and abrasion.

Scrubber — Wild cattle which live permanently in thick scrub country.

Scrub-runner — A stockman who specialised in galloping through heavy scrub, and catching and throwing scrubbers which otherwise could not be mustered.

Scruffing — The process of catching calves in the stock yards and throwing them on their side preparatory to marking and branding.

Sesamoid bones — Two small bones at the rear and outside of a horse's pasterns, which form part of the weight-bearing and shock-absorbing system.

Shin tapper — A hobble chain strapped to one pastern only, which stopped horses travelling too fast. That is, excessive speed caused the chain to swing wildly and continually 'tap' the opposite pastern and cannon bone.

Sideline — A length of chain strapped to front and rear pasterns on the near or offside. Used mainly on horses which wandered excessively. In conjunction with hobbles, the animal had three legs fettered and was considerably restricted in its movements.

Skewbald — Generally refers to horses which have brown and white colouration.

Slick forks — Early USA saddles did not have an equivalent to Australian kneepads and thus the fork (the cantle) was 'slick' (smooth.)

Slicker — Waterproof coat.

Snaffle bit — A bit commonly used in Australia. Unlike solid bar bits, snaffle bits are comprised of two pieces joined in the middle.

Snig — To pull the animal closer and continually take up the slack in the rope.

Suffolk Punch — A light draught horse breed which originated in East Anglia in the early 1500s.

Surcingle — A leather strap which wraps completely around an animal and its associated saddlery and which, when tightened, better secures the load.

Suspensory ligaments — ligaments which run down the cannon bone, over the sesamoid bones and are fixed to the internal hoof structure. They form part of the weight-carrying and shock-absorbing mechanism of a horse's leg.

Tack — Harness.
Tail/tailer/ tail-out — A horse tailer is responsible for the care and well-being of a plant. He has to ensure his charges are placed on adequate feed at night, and that they are mustered and in ready for work before dawn. Thus he 'tails' the team out.

Tapadero — A leather facing which covers the stirrup (USA). Often lined with sheepskin, they provide warmth and protection. Also have a decorative purpose.

Top End — The Territory coastal plain.

Trepang — Fishermen who catch the sea-slug trepang, or Beche de Mer.

Twitch — An eighteen inch length of wood with a short loop of thin rope or twine at one end. The loop is placed over a horse's nose and turned until it is quite tight. The resultant pain tends to make the horse stand still while it is being treated.

Vesta — a wax-coated, waterproof match.

Whip horse — A horse used specially for pulling water up from wells. (ie. 'whipping it up.')

Winkers — Leather harness designed to occlude a horse's lateral vision and thus force it to concentrate on the task in hand.

Wither — The highest part of a horse's back, which is located at the junction of the back and neck.

Index

abattoirs 78
Aboriginal
    depredations 273
    labour 111
    musterers 51, 52, 53
    skills 51, 52, 53
    stockmen 13, 34, *51*, 146, 158, 181, 182, 267
    trackers 268
    troopers 266, 267
        executed 266
    women 173, 256, 269
        sexually abused 267
Aborigines 15, 16, 18, 29, 56, 78, 208, 210, 286
    armed 160
    award wages 13, 30
    burning practices 303
    dispossession 251
    hostile 10
    killing pastoral stock 251
    murder of 266
    murdered by Fraser 253
    Queensland 266
    spearing cattle 29
    station 32, 199, 256, 258
    tribal 252, 253, 256, 267
    warfare against settlers 250-272
anhydrosis 165-166
ants 84, 330
    white 281, 282, 291, 294
Arabian horses 123, 125, 127-31
    stud 129
    unpopular 130, 131
artesian water 308-316
    first bore sunk 309, 310

Arundels 169, 181
Auld, W P 140
Australia, crossings , 79, 140, 142
Australian Light Horse 130, 141, 320

Baker, Reg 51, 69, 130, 207
Banks, Sir Joseph 263
Bannister, Thomas 188
barbed wire fencing 30, 77, 295, 296
Barker, H M 12, 14, 26, 45, 53, 54, 81, 105, 111, 131, 135, 155, 202, 207, 208, 220, 221, 248
    comments on camels 226, 227, 228, 229, 230, 231, 236, 237, 238, 239
    cueing bullocks 157-158
    wagon purchase 247
Barkley Tableland 23, 79, 101, 134, 135
Bartley, Nehemiah 23, 96, 304, 305
Bates, Don 171
Bates, George 135, 136, 169, 173, 178, 179, 180, 181
Beckler, Hermann 226, 227
Bedourie camp ovens 105, 329
beef 24
    consumption 24
    eaten by drovers 101
    factories 75
    herd 28
    industry 75, 78, 79, 81, 112
        Leichardt-Gilbert region report 68
bells 98, 324, 325, 326
    Condamine 324
    for wandering stock 161, 162
billies 326
Bischoff, Edwin 324

Bischoff, Stan 39, 41, 44–50, 54, 56, 70, 72, 134, 324
    injuries 49
    mustering camp *54*
    skills learnt from Aboriginal musterers 51-53
blacksmiths 245, 302
Blue Mountains 252, 318
    crossed by settlers 18, 168
Boer War 141, 173
Boldrewood, Rolf 143
bore-drain delvers *314*
bores 35, 79, 94, 112, 241
    artesian 308-316
    drains, equipment *312*
    drilling 309-312, 315
Boucaut, John 129
Boxall, George 16, 263
Bradshaw, Robert 37, 38, 47, 130
Brady, Jack 202, 207, 208, 213, 214
branding cattle 58, 61, 62, 63, 68-69
    broncoing technique 63-68
Brigalow Bill 253, 254
Bright Brothers 171
Britain 110, 160
    houses imported from 278
    spencilling technique 160
British
    horsebreakers 196-199, 201
    horses 114, 117, 121, 127, 139, 196-197, 198
    migrants 117, 176, 289
    pack-saddles 188
bronco branding *66*
    calves *67*
    cattle 63-68
    horse *64*
    ropes 320, 321-322
    yards 292
Bruce, Alex 120, 122
brumbies 200, 210
    mustering 69-70
BTEC scheme 13
    eradication scheme 133
buckjump riders 199
buckjumping 201, 212, 219
buffalo 76, 107, 164
bull-hide leggings 51
Bullita station 100, 269
bullock
    cue *157*

    drawn wagons 103, 284
bullocks
    cueing 157-158
    dehorning 43, 44, *62*
    draught animals 224-230, 237, 238, 239, 240, 241
    dry-daying 94
    'fats' 156
    footsore 150, 156, 158
        husbandry 240, 241
    hide, used for saddles 180
    mainstay of cartage industry 224, 225
    sorting and marking 56-58
    wandering 158, 159, 227
bulls
    old and of little value 55, 56
    shooting of old 55, 56
    young, thrown, castrated and earmarked 55
bullwhips 237
burning of land 35, 296, 298-304
bush fires 296, 303, 304
bush life 21, 73, 160, 317
    equipment and know-how 317-331
bushmen 51, 96, 97, 107, 122, 146, 151, 160, 161
Butler Brothers 187
Butler, Lance 185, 186
Butlers, Edward 180, 182, 185

Caley, George 263
calf cart 85
calves 27, 58
    branding and marking 58, 61-63, 68-69
    bronco branding 63, 66, 67-68
    droving 84-85
    roping 66, 68-69
camels 96, 225-232, 236, 237, 238, 239, 240, 249, 284
camp horses 57
campfires 87
camping, dangers 272
camps 33, 34, 105, 136
    mustering and droving 99
    skeleton 34
campsites 86-87, 239
Canning route 81, 94, 96
Cannon, Arthur 231, 233, 237
Cape of Good Hope 16, 113, 114, 126

ponies 126
Carnarvon Ranges 69, 70
cartage industry 224, 246
cattle
    branding and marking 58, 61-69
    bronco branding 63-68
    dehorning 43, 44, 45, 56
    dips 108-110, 292
        Queensland government-
        approved *108*
    drafting *57* , 59
    droving 10, 18, 20, 21, 23
    finding skills 33, 34-35, 37-38
    first in colony 16-17
    from Cape Town 16
    herds 28
    holdings see stations
    horses 123-124, 130, 131, 132
    industry *see* pastoral cattle
    industry
    injuries and death 70, 90, 93
    mobs
        counting at handover to
        drovers *86*
        optimum size 83, 95
        rushes 87-93
        small 28, 34
        watering 94, 95, 96
    mustering 17, 18, 27-72
    overlanded to Adelaide 20, 81, 82
    pleuropneumonia 110-111
    prices 24, 75, 77
    quiet 37, 56
        see also coachers
    scrubbers 44-47, 53-56
    scruffing *61*
    shifted from New South Wales
    to South Australia 20, 21
    shipped to Philippines 27
    sorting *59*
        and marking 56-69
    spearing 29, 268
    swimming across rivers 28, 97, 98
    throwing and tying 40-44, 52-53
    tick 29, 107
    trade with South East Asian
    countries 23, 24
    watching 88
    watering 83-84, 94-95, 96, 97
    wild 16, 17, 18, 32, 37-39, 44-47, 52
    working skills 11, 37
    yarding 58, 59, 87
    yards 289-296
        fork-and-rail 289
        paperbark *290*
        plan *293*
        post-and-rail 289, *291*, *292*
        rail-and-sleeper *295*
cattlemen 78, 79, 131, 132, 165, 166, 172, 200, 264, 286, 294, 295, 298, 299
    Kimberley 78, 181
    Northern Territory 79
    Queensland 306
    USA 132
Central Australia 150, 162
chain manufacturing 168-169
Chinaman's lane 92
Chinese drovers 88
Chinese labourers 283, 306
Christison, Robert 127
clothing 32, 317-320
Clydesdale horses 134
coachers, defined 37
Cockburn-Campbell, Thomas 36, 89, 200, 204, 206, 208, 209, 218, 237, 328
    mud bricks technique 279-280
Cole, Tom 14, 100, 103, 212, 218, 222
Colony of New South Wales 16, 113, 114, 168, 175
    colts bred in 175
    first verandah 277
Condon, Ron 207, 269
convicts 16, 17, 19, 113, 263, 318
cooking equipment 328-329
Coolgardie safe 289
cow ponies 132
cowboys, American 9, 10, 11, 30, 31, 40, 42, 58, 68, 92, 99, 132, 143, 179
    boots 319
    compared with stockmen 9-11
    leather gloves 320
    saddle-bags 328
    saddlery 192-195
cowkickers 206, 207
Cowpastures region 16, 17
cows 27, 58
crocodiles 35, 166
Cunningham, Peter 14, 318
Curr, Edward M 118, 120, 122, 123, 125, 127, 128, 130, 173, 176

Dale's Tannery 180

372

Darling Downs, first station 21
Dean, Frank 172, 284, 287, 296, 307
Delamere station 57, 58, 94, *274*
Delouer, Ned 234, 235
Depression 151, 169
diseases 29, 107, 110, 111, 165, 166
    gravel 164
    redwater fever 78, 107
distance riding 118, 122, 128, 139, 140, 145
dogs 12, 13, 33, 45, 100, 233, 253, 258, 259, 268, 269, 270
    native 24, 25
    role in pastoral cattle industry 37-38
donkeys 221, 225, 226, 234–36, 249
Dorisvale station 133
Dowling, Joe 41
draught animals 224-249
drought *see under* weather
drovers 111, 123, 136, 156, 158, 165, 222, 280
    American 10, 103
    Australian 10, 12, 103
    bells 325
    blinded by dust 74
    camp *83*, 84
    campsites 86-87
    Chinese 88
    equipment *102*
        transportation problems 101-107
    first 19, 20
    first major droving trip 81-82
    food and equipment 101-107
    hardships 105
    Pilbara 36, 200
    pioneer 19, 20, 81, 94
    skills 19, 20, 21, 84, 85, 89, 93
    techniques 82-85
drover's plant *82*, 84
droving 14, 73-112
    in Australia and USA 74-78
    cows and calves 84-85
    described 74
    end of 14, 112
    first major trip 20, 81-82
    horses 111-112
    large-scale 81, 94
    livestock 18-21, 23
    problems 85

sheep 19, 20, 21, 26
dry season 27, 28, 136, 240, 299, 330
Duracks 25
dust 74, 75
Edgar, Felix 29, 34, 84, 87, 259, 269
Edward Butlers 180, 182, 185, 186, 187
Edwards, Ron 317
Electra plane crash 234
Ellendale homestead 277, 278
Elsey station 32, 55, 58, 62, 79, 88, 201, 242, 260, 283
Emmanuels 34
England 117, 136, 139, 174
    horses imported from 114, 127, 133
equipment 317-331
    drovers 103-105
    transportation problems 101-107
Evans, George 18, 144, 145, 158
expeditions 123, 151, 161, 162, 163, 166
    De Grey to Port Darwin 25, 165, 193
    Leichardt 191
    North Australian 22, 25, 140, 145, 146, 149, 151, 162, 167, 189
    North-West 149, 151, 163
    Port Essington 156
    Western Australia 167, 188
explorations 147, 150, 158
    Blue Mountains 252, 318
explorers 22, 103, 122, 123, 140, 144, 145, 146, 151, 158, 161, 166, 228
    contribution to pastoral cattle industry 18, 19, 21
Eyre, Edward John 19-20, 81, 123, 140, 146, 163, 188, 239

feed 27, 79, 86, 93, 98, 117, 118, 120, 123, 131, 136, 138, 158, 225, 300, 301, 307
    spinifex and scrub 135
fenced stock routes 84, 87
fences 273, 295-296
    barbed wire 30, 77, 295-296
fencing
    important in rangeland management 316
    of prairies 30, 36, 77
    of stations 30, 36
    steel posts 296
    wooden posts 295
firearms 250-272
    improvements 264-265
    inefficiencies 262-265

muzzle-loaders 262, 263
protective practices 259, 260
revolvers 265-266
rifles 54, 255, 265, 266
firebreaks 302
fireplough *301*
fires 35, 296, 298-304
camp 87
firewood 101
for steam engines 315, 316
Fisher (pioneer) 25
Fleming, Jim 133, 287, 291
Fleming, Mick 133
Fletcher Moore, George 122
flies 39, 71, 75, 84, 126, 147, 330
blow 26
Flora Valley station 205, 206
Fogarty, Ted 57, 58
food 19, 20
and equipment, difficulties conveying 101-107
see also beef; meat
Forrest, Alexander 25, 165, 191, 193
Forrest brothers 122, 123
Fossil Downs 284, 286, 288, 289, 300
Fraser family massacre 253
Fraser, Simon 309, 310
Fraser, William 253
Furphy, Joseph 10, 97, 99, 170, 174, 212, 215, 218, 220, 233

Gabb, Andrew 171
Gardiner, Jack 184, 185
Gibson desert 162
Gibson, James 23
gidyea (Acacia cambagei) 303
Giles, Alfred 94, 140, 164
Giles, Ernest 107, 122, 123, 162, 228, 265
Gliddon, Joseph F 295-296
Gordon, Adam Lindsay 217
Gordon, Coralie 129
Gordons 25
Gould, Nat 139
Government 79, 129, 309
bores and artesian water issues 310, 311, 312
legislation 298, 314, 315
Queensland 24, 307, 309, 314
grasses 35, 118, 126, 299, 300
bladey 145
burning 299-304

cane 287
kangaroo 35, 297
spear 25
grasslands 18, 22, 25, 101, 117, 140, 301
management 298
Gray, Bob 264
graziers 268, 308, 309
pioneer 18, 74, 176
sheep changed for cattle 26
grazing 22, 297
damage to rangelands 297-298
Gregory, Augustus 22, 25, 140, 145, 148, 151, 162, 167, 189-191
Gregory brothers 140, 148, 149, 167
Gregory, Francis 148, 151, 160, 163, 164, 297
Gregory pack-saddles 189-192
Grey, George, Governor 21
Gulf of Carpentaria 161, 162
properties, abandoned 25
region 24, 63, 81, 166, 191
settlers 23, 24
Gunn, Aeneas 55
Gunn, Jeannie 14, 32, 55, 58, 88, 201, 240, 242, 260, 281, 283, 287
guns see firearms

hamburger meat 56
harness 173, 230, 231
maintenance 149
harness-makers 168, 169
Harris, Alexander 117, 118, 160
Hawdon, Joseph 20, 81, 82
Haygarth, William 14, 174
Henry, Ernest 176, 231, 289
Henwood, Annette 284, 285, 286, 289, 300
Heytesbury Pastoral Group 14
Hill, Jim 178, 181, 183, 186, 187
Hill, Syd 185, 186
hobbles 32, 33, *159*, 169, 170, 202, 204, 207, 208, 209, 226, 227, 320
hobbling of stock 159, 160, 161, 223, 324
homesteads 268, 269, *275*, *276*, 279, 284, 286, 287
brick 286, 287, 288, 289
bough sheds 289
Cairdbeign, 281 (Plate 62) 283
construction and safety factors 271-272
Dorisvale 279

first buildings 273
Fossil Downs 284, 286, 288, 289
insulation 287
Nissen-hut-style *285*
Rainworth 271
roofs 282-283, 284
sheds 285, 286, 289
stockyards 289-296
stone buildings 279, 284
Strathdarr 280, 287
thatched-roofed meat house *288*
timber 280, 281, 282, 284
verandahs 277, 278, 283, 286, 287
white ants damage 281, 282
yards, and fences 273-296
Hoover, Herbert 248, 249
horse herd, growth of 113-117, 121, 129
horse husbandry 123, 125, 140, 141
horse plants 97, 98, 99, 100, 101
horse-tailers 83, 84, 98, 99, 103, 325
horsebreakers 51, 168, 178, 194, 200, 207, 208, 209, 219
    boots 318
    British 196-197, 198, 199, 201, 209
    deaths 199
    emotions 218-219
    riding style 217-218
    saddle *177*
    skilled 200, 201, 202, 207, 213
    USA 198, 201, 213, 214
horsebreaking 128, 132, 196-223
    deaths 221
    mouthing 205, 207, 208, 220
    pack-saddles 221-223
    saddling 209-211
    without yards method 220-221
    yards 292, 294
horsehair, for saddles 171-173, 183
horsemanship 10-13, 37, 123, 138-167
    in decline 140
    defined 138-139
horsemen 200, 211, 217
    descended from British stock 194
    fear and nervousness 218-220
    skills, lost 12-13
    USA 192
horses 13, 67
    ailments and diseases 164-167
    American 131, 132, 135
    Arabian 114, 123, 125, 128, 129, 130, 131

bad 100, 128, 217
bells 161-162
best bronco 133
best mustering 133
bits 194, 197, 208, 209
bought for mustering teams 29-30
breeding 113-137
    Australian style 126
    problems 125-127
breeding conditions 125, 126
breeds 126-133, 134, 135
British 114, 117, 121, 127, 139, 196-197, 198
bronco *63*, *64*, 133
brumbies 172, 200
    mustering 69-70
bucking 174, 175, 176, 178, 179, 181, 182, 201, 212, 215, 217, 218, 219, 220, 222
camp 33, 57-58
Cape 113, 114, 120, 126
care 138-150
catching 99, 100, 101
    for cattle work 123-124, 130, 131, 132, 220
cavalry 118, 120
clumpers 133
Clydesdale 134
Clydesdale-cross 97
colonial 113, 114, 117
deaths 141, 165, 166, 167, 208, 238
dehydrated 162-164
deterioration 119, 120, 121
draught animals 224-230, 236, 237, 238, 239, 240, 249
droving 111-112
early explorers 147-149
endurance 118, 122, 123
    races 128
exported to India 118, 119, 120, 122, 127, 201
finding 98-99
fitted with bull-hide leggings 51
footsore 150, 153
from Cape of Good Hope 113
galling 141, 144, 145, 173, 188
good quality 123, 124, 125, 126, 133
hardiness 117, 118, 120, 123, 131
hobbling 159, 160, 161
imported from
    Britain 114, 127, 133

       Cape or India 114
       Lombock and Timor 114
       stock 113, 114
  injuries 50-51, 70, 141
  kicking 206, 207
  lack of 29
  limitations 138-140, 162, 163
  mustering 32, 33, 48, 49, 69-70, 133
  packhorses see packhorses
  Percherons 132, 133, 134
  prices 114, 117, 127, 130
  quarter 135
  race 119, 121, 127, 128, 166
  rearing 125, 126
  saddle 118, 119, 120, 121, 134
      deterioration of 119, 120, 121
  scrub-runners 48-50
  shoeing 150-156
  shortage 29, 30
  shot 100, 133, 147
  sore backs from saddles 141,
     173, 181
  stamina 117, 118, 123
  station 133, 198, 199, 201, 215
  statistics 114, 117
  stock 33, 123, 124, 125, 127, 128, 132
  sufferings 50, 147, 148, 149, 150,
     162, 163
  Suffolk Punch 133
  swimming in rivers 97-98
  teeth problems 208-209
  temperament 142
  thoroughbred types 127, 128,
     132, 133, 134, 215
  thoroughbreds 48, 119, 121,
     127, 130, 131, 133, 134
  trotting 142, 143
  wandering 159, 160, 161, 227
  whip 95, 96
  wild 131, 199
  working 63, 118, 130, 131, 150
  see also nighthorses
housing
  colonial 277
  kit homes 278
  prefabricated wooden homes
     278, *279*
  see also homesteads
Hoy, Jim 9, 10, 11, 319
Humbert River station 208, 257

Iddon, Ron 120, 121, 122
Idriess, Ion 14, 29, 103, 130, 252, 260
India 24, 114, 118, 120
  Australian horses in 118, 119,
     120, 122, 127, 201
  horses imported from 128

Jack, Humbert *43*, 44
Jack the Quiet Stockman 201
Jack, Robert Logan 309
Jack, Werabone 50, 51, 52
Jardine brothers 166
Jones, Samuel 324, 325

Kallara station, first artesian bore
  sunk 309
kangaroo hide whips 322, 323
Keggabilla, photo of muster *42*
Kelley, J H 68, 299, 303
Kimberley 27, 29, 30, 34, 78, 79, 134,
  135, 166, 191, 228
  cattlemen 79, 81
  country unsuitable for sheep 25, 26
  east 225, 226
  mustering problems 29
  properties 27, 34, 134
  stations 79, 134
  steel posts 296
  stock routes 79
  walkabout disease 111
  west 26, 29, 34
King, Governor 17, 18
Kitchener, Lord 173
Kite, Bert *210*, *214*

La Meslee, Edmond 242, 243, 244
land
  arid regions 78, 125, 135, 140,
     225, 228, 247, 299, 308, 311
  blacksoil plains 286, 300
  boggy country 71-72
  burning 35, 296, 298-304
  coastal 126
  coastal plains 35, 71, 72
  deforestation and degradation
     315-316
  desert country 163, 316
  limestone country 32, 35, 151
  marshy country 35, 148
  open country 33, 36
  overgrazing 26, 98, 297, 298

plains country 22, 24, 31, 33, 34
prickly pear country 45, 47, 48
rangeland management 297-316
scrub see scrub land
spinifex country 297, 299, 301
swamps 35, 145
unsuitable for sheep 25, 26
water indicators 93-94
wooded 37, 38
see also grasslands
Landsborough, William 191
leather 194
    curing and tanning 320
    greenhide 320, 321, 322
    kneepads 176-179
    for saddles 168, 178, 180
Ledger, Dave 34, 37, 100, 130, 131, *154*
Leichardt, Ludwig 22, 79, 93, 101, 135, 142, 146, 150, 156, 161, 191, 262, 321
    horse deaths 166
Leichardt pine 284
Leslie, Patrick 21
liver fluke 25
livestock see stock
Loos, Noel 23, 266, 268
Loughead, J S 309
Lyons (pioneer) 25

McCoy, Joseph 76
McDonald, Reg 284, 286, 288
McDougall Stuart, John 79
Macgeorge, A J 96, 238
Makin, Jock 28
massacres 253, 268, 271, 276
meat
    curing 329
    houses 287, *288*
    sun-dried 329-330
    see also beef
Mitchell
    Sir James 169
    Surveyor-General 103, 239, 253
moonlighting 53-55
Moore, Joe 280
Mora, Jo 30, 36, 40, 58, 103, 131, 132, 179, 187, 191, 193, 195, 198, 219, 220, 265, 319
Morton, W 260-262
motor vehicles 224, 246, 248, 249
    cars 246, 249
    trucks 235, 236, 249

Mueller, Ferdinand von 140, 167
mules 135, 136, 204, 222, 225, 330
Murraculcul station *54*
Murranji track 79, 91, 92, 94, 112, 228
musterers 46
    Aboriginal 51, 52, 53
    scrub 48
mustering 27-72, 140
    brumbies 69-70
    camps 28, 33, *54*, 99, 200, 212, 213, 221
    conditions 27, 28
    described on Elsie Station 32-34
    difficulties 28-30, 32
    horses 48, 49, 69-70, 133
    in USA 30-31, 36
    moonlighting 53-55
    plant 100
    poor practices 199
    quiet cattle 37, 56
    skills 17, 18, 37, 38, 39
    techniques 33-35
    using drums 17
    wet-season 27, 28
    wild cattle 17-18, 37-56
        different methods 37, 44, 53, 55
        moonlighting method 53
musters 13
    clean 29, 30
    collective 36

Native Police 266, 267, 268
Nemath, Brian 322, 323
nighthorses 87, 88, 89, 90, 91
nightwatchmen 82, 87, 89, 92
Noble, Jack 100, 136, 137
North Americans, payments for bulls and bullocks 56
North Queensland 29, 299, 306
    discoveries 22
    settlement 22
northern pastoral cattle industry
    deterioration of 13
    expansion of 21-26
    life 11-12
    technological changes 12-15
    and USA pastoral history 9-11
    see also pastoral cattle industry
northern pastoral industry
    beginnings and expansion 16-26
    rescued by gold discovery 24

northern ports 23
    trade with South-East Asian countries 23-24
Northern Territory 29, 30, 63, 69, 78, 79, 100, 107, 112, 134, 225, 228, 302
    artesian water pumps 315
    cattle ban 78
    field surveys of station boundaries 30
    first property stocked 23
    sheep problems 25
    stock routes 79-80
    stocking of properties 79
Northern Territory Times 180, 225, 265

Oenpelli Anglican mission 134
outback 10, 73, 165, 264
    experience 11
overland journeys 19-21, 81
overlanders 20, 128
    clothing 317
    described 21
Oxley, John 18, 144, 147, 148, 149

pack animals 221-223
pack-saddles 187-191, 221-223, 326
packhorse drovers 103-106
packhorses 21, 103, 104, 105, 133, 147, 147-150, 162, *193*, 207, 328
    overloading 147-149
Palmer, Edward 14, 22, 74, 124, 224, 246, 306, 308, 311, 318
pastoral cattle industry 9-15, 78
    beginnings and expansion 16-26, 78
    horses role in 113
    profitability 78
    USA 9-11, 74-78
    see also northern pastoral cattle industry
pastoral exploration 18-25
pastoral properties 10, 13, 19, 78, 134, 135, 309
    abandoned 78
    Darling Downs region 21
    destocked and abandoned 25
    purchased and restocked 25
    establishment 18-26
    fattening, western Queensland 81
    management 297-316, 298, 299
    unfenced 30, 31
    Vesteys 27

    see also stations
pastoral regions 23
    stocking of 23, 24, 25, 251
pastoral runs 18, 23, 121
    Gulf region 23, 24, 25
pastoral settlement 21-26
    state of Queensland 21
Paterson, Banjo 69, 121, 126, 141, 142, 173, 211, 212, 305, 313
Pearce, T H 129
Peel River Land and Mineral Company 134
Percheron horses 132, 133, 134
pigskin 181
Pilbara 74, 135, 151, 225, 226, 287
    properties 36, 273
    red soil 279, 280
pioneers 18, 25, 166, 176, 251, 287
    drovers 19, 20, 81, 94
    North Queensland 124, 264
    Queensland 22, 74, 224
    settlers 21, 22, 23
    skills 18-19
    west Kimberley 26, 252
Pit Waddel 168
plants
    poisonous 29, 51, 135, 165, 166, 167
    woody weed 303
pleuropneumonia 107, 110, 111
ploughs
    fire 301, 302
    mouldboard 306
poisonous plants 29, 51, 135, 165, 166, 167
police
    mounted 107, 318
    native 266, 267, 268
    patrols 29, 252
ponies 114, 126, 131, 132, 189
    Cape 126
    Timor 114, 133
Poole, Peter 10
population 78, 114, 170
    horse 114
    rural 332
    Taroom region 250
    USA 75
Port Phillip settlement 20, 81
prickly pear country 45, 47, 48, 51
puffs 165-166

Queensland 29, 30, 69, 81, 84, 107, 302, 306, 307

Aborigines 266
  bores sunk 309-311
  Carnarvon Ranges 69, 70
  coastal 63
    infestations of liverfluke, worms, spear grass 25
  exploration and settlement 21-26
  fencing 316
  first station, on Darling Downs 21
  Government 307, 309, 314
  moonlighting practice 53-54
  pastoral settlement 21
  pioneers 22
  saddlers 178, 187
  stock numbers 25
  stock routes 81, 87
    gazetted 79, 98
Queensland Government 24, 310, 311, 314
  recommended cattle dips 108
Queenslander, house style 277
Quilty, Tom 181

racehorses 119, 121, 127, 128
  exported to South-East Asia 166
racing
  endurance 128
  thoroughbred 121
racing industry 126, 127
rail transport 75, 76, 77, 78, 79, 81
railway sleepers 294, 295
railways 225, 284
Rainworth homestead, *271*, 283
ranching, traditions 10
rangeland management 297-316
rangelands
  deforestation 315, 316
  degraded by overgrazing 297-298
  woody weed 303
Rangelands Issues Paper 298
Ranken, W H L 185
redwater fever 78
Reid 270
rendering works 78
Reynolds, Henry 51, 250, 256
Richardson, Henry Handel 139
riding boots 317-318
rivers 22, 23, 81, 93, 151, 165, 166, 286, 287, 289, 297, 308
  crossings 97-98, 101, 102, 103
  Daly, Victoria and Ord 25

road making, Broome 235
road transport 112, 117
roads
  development 224
  poor 170, 242
Robinson, E O 260
Rocklands station 23
rodeos
  Australian 9
  USA 9
Roe, John Septimus 188
roping techniques 63, 66, 68, 69
Ross, Peter 158
Royston, General 211, 212
rushes, prevention of 85, 89, 90, 91, 92, 93

saddle trees 170-171
saddle-bags 105, 107, 326-328
saddle-blankets 145
saddle-horses 118-121, 134
saddlers 145, 168, 178, 184, 186, 187, 317
  Arundels 181
  station 170, 172
  travelling 170
  USA 192, 193
saddlery 168-195
  adornment 194-195
  businesses 168, 169, 170, 180, 181, 182, 184, 185, 186, 187
  regional differences 191
  USA 191-195
saddles 10, 11, 63, 68, 132, 136, 145, 149, 209
  American 10, 63
  army 174
  Barcoo Poley 183
  'Black Boys' 182
  British 173, 174, 188
  cavalry 174, 188
  English hacking 176
  English hunting 173-6
  kneepads 176-181
  military 173, 188
  pack 187-191
  pastoral 194
  prices 182, 186
  Quilty 181, 182
  race 171
  riding 187, 188, 191
    replaced by pack 191

show 171
   stock 10, 63, 170, 171, 181
   Wieneke *177, 180*
Savage, Matt 12, 14, 31, 165
sawmilling 284-285
Schmidt, Dolf 134
Schmidt, Felix 111, 316
Schneider 185
Schultz, Charlie 40, 41, 42, 56, 58, 67, 91, 92, 95, 103, 158, 202, 203, 208, 213, 254, 257, 282, 284
Scott brothers 268
Scott, Joy 187
scrub bulls
   dehorning *44*
   mustering dangers and problems   55-56, 72
   shooting 56
   used for hamburger meat 56
scrub dashers 44-45
scrub land 17, 18, 33, 36, 37, 44, 45, 85, 86, 90, 91, 93, 135, 326
   Mungle area 45
   Murranji 79, 91, 92
scrub riders 52, 53
scrub runners 48-50, 130, 184
scrub running 48-50, 52, 134
scrubbers 44-47, 53-56
   defined 44
   moonlighting practice 53-54
   stalked and shot 54
Searcy, Alfred 14, 252, 260, 262, 272, 282, 325, 330
settlement 160, 224, 225
   of Australia, killings 250
   conflict between Aborigines and settlers 250-272
   expansion of 21-26
   of North Queensland 22
   of Northern Australia 21-26
   of Peak Downs district 22
   Port Dundas 262
   Raffles Bay 107
   Western Australia 169
settlers 119, 167, 168, 253
   early stock 22
   Gulf 23
   killed by Aborigines 250, 253, 254, 260, 262
   pioneer 21, 22, 23
sheds 285, 286, 289

sheep
   droving 19, 20, 21, 23, 26, 94
   overlanding 19, 20, 21, 81
   replaced by cattle 26
   stock rise in Queensland 25
   unsuited to country 25-26
shipping 77, 81, 284
   costs 114
   horses 122
   supplies for stations 101
shoeing stock 150-158
   bullocks 156-158
   horses *154, 156*
sideline/sidelining 160, *161*, 202
Simpson, Bruce 14, 41, 57, 58, 83, 87, 97, 100, 105, 123, 124, 128, 130, 155, 207, 217
Sims, Jack 187
Siringo, Chas A 30, 31, 72, 92, 93
Skuthorpe, Archie 199
Skuthorpe, Dick 199
Skuthorpe, Lance 199
soils 103, 117, 167, 313
   blackplain 286, 300
South Africa 110, 120, 141, 173
South Australia 20, 21, 29, 81, 129, 168
South Australian Government 122, 129
South-East Asia 23, 166
Spanish horses 131
spencilling stock 160
spinifix country 135, 297, 301, 303
Springvale station 94, 164, 281
spurs 328
stampedes 85, 87, 89-93
State Shipping Service 284, 285
stations 28, 29, 30, 32, 94, 133, 134, 135, 136, 164, 176, 273, 280, 281, 282, 283, 289, 295
   Bullita 100
   Burnside 27, 28, 29, 38, 107, 136, 145, 299
   cattle dips 108, 109
   cattle numbers 23, 28, 299, 300
      carrying capacity 299, 300
   destocking 25, 298
   difficulties pioneering 29
   fences 295-296
   fencing 316
   first on Darling Downs 21
   Katherine district 94, 164
   Kimberleys 79, 111, 316
   Murrakai 27, 28, 29, 35, 299

northern 111, 164, 165, 183
Queensland, subdivision 300
size 28
stocked 23
stocking rates 21, 22, 23, 25, 28, 299, 300
Taroom district 37
Territory 316
unfenced 30, 31
Willeroo 129
yards 62-66, 68-70, 273, 289-96
    branding equipment 68
    construction 58, 62, 63
    with races 62, 63
see also names of; pastoral properties
statistics
    cattle 28
    horses 114, 117
    killings during settlement and occupation 250
    population 75, 78
stock
    ailments and diseases 164-167
    control techniques 17, 18, 21
    large numbers moved 20-21, 28
    overlanding 19, 20, 20-21
    overworking 146, 147-149
    restraining methods 160-162
    shifted during dry years 307
    spearing 29, 268
    swimming 97, 98
    thirsty 94, 95, 96, 97, 162, 163
    unmarked and unbranded 32
    wandering 158, 159, 160
    watering 307, 312, 313
    working 158, 307
stock horses 33, 118, 123, 124, 125, 127, 128, 132
    breeding 125, 126, 127, 128
    good characteristics 123-124, 127-128
    Waler type 128
stock inspectors 107
stock routes 79, 80, 94, 98, 112, 307
    Barkly 104
    dams and tanks constructed 307
    fenced 84, 87, 98
    Northern Territory 79-81
    Queensland 79, 81, 87, 98
stock saddles 168, 170, 180, 181, 184, 186, 187, 192

described 185
development 184-187
father of, dispute 184, 185, 186, 187
stockmen 12, 184, 200, 201, 255, 280, 296
    Aboriginal 13, 34, 51, 146, 158, 181, 182, 267
    bells 325
    clothing 32, 317-320
    compared with cowboys 9-11
    dangerous conditions 32, 71-72
    emotions 218
    equipment and know-how 317-335
    hardships 71-72
    hemorrhoids 180
    leggings 318, 320
    riding boots 317-318
    riding style 217-218
    saddles 180, 181, 185
    scrub-dashers 44-45
    shortcomings 17, 18
    skills 12-14, 32, 37, 41, 57, 58, 70, 71, 90, 91, 97, 138
    spurs 328
    station 12, 13, 32
    supplies and equipment 21
    throwing cattle 40-43
    working in yards 58-63
stockwhips 33, 40, 202, 210, 222, 320, 322-324
stockyards 32, 289-296
    bronco yards 65, 66
    construction 58, 60, 62, 63
    crude, for brumbies 69
    wire 58, 290-291
Stokes, John Lort 21, 22
Stuart, Charles 18, 122, 123, 140, 141, 142, 163
Sturt, Charles 81, 141, 142, 150, 162, 253, 328
Sutherland, George 23, 26
Syd Hill Saddlery 169, 170, 186

tanneries 56, 168, 180
Thacker, W 127, 128, 130, 201
Thompson, Bob 252
thoroughbred horses *see under* horses
thoroughbred racing 121, 126, 127
throwing cattle, techniques 40-43, 52-53
Ti-Tree country 71, 287
Tietkins, W H 265
Tilmouth, H 260

381

timber 282, 284, 313
> dwellings 280, 281
> for fences 295. 296
> for wheels, 247
> for yards, 289, 290, 291
> *see also* wood

Timor ponies 113, 133
tobacco 330-331
Tommey Tommey, Jackey 52
Toohey and Thornton 247
Top End region 35, 71, 133, 145, 166, 299, 303
Tothill and Gillies study 297
trackers 51, 268
tracking skills 37, 160, 161
transport 117, 249, 284
> buggies 245, 246, 249
> bullock-drawn wagons 103
> cartage industry 224, 246
> draught animals 224-249
> high costs 26, 29, 198, 295, 296
> horse-drawn carriages 242, 243, 246
> motor cars, 249, 246
> motorised 138, 246, 248
> rail 78, 79, 81
> railroads in USA 75, 76, 77, 78
> shipping 78, 81
> sulkies 242
> traps 245
> trucking of cattle 112
> trucks 249
> vehicles, design and maintenance, 241-247
> wagons *see* wagons

travelling saddlers 170
Trew, Pompey 329
trucks 112, 249
Tuckfield, Francis 251

Underwood, Pat 29, 135, 200, 207, 302
United States of America 63, 92, 103, 131, 257, 310
> barbed wire fencing 295-296
> cowboys *see* cowboys
> droving 74-78
> horsebreakers 198, 201, 213, 214
> horses 131, 132, 135
> houses imported from 278
> kneepads 179
> mustering 70
> pastoral cattle industry 9-11, 30, 36

> population 75
> saddlery 191-195
> saddles 179, 180

Van Diemen's Land 175
vehicles *see* motor vehicles; transport
Vesteys 27, 107
Victoria River Downs station 28, 31, 32, 59, 62, 63, 64, 65, 211
> camp scene *83*

wages 13, 30, 114, 130, 178
wagonette horses 133
wagons 19, 103, 238, 241, 246, 247, 248, 284, 285, 308
> Chuck 103
> bush wagonette *245*
> drays 21, 307
walkabout disease 29, 111, 165, 166
Walker, Frederick 266
Ward, Billy 253
water 224, 225, 286
> artesian 308-316
> bores 79, 94, 95, 241, 308-316
> horses doing without 123
> shortages 20, 97, 308
> sources 33, 34, 35, 36, 79, 93, 94, 95, 96, 97, 112, 228
> > fouled by cattle 75
> supplies 162, 163, 303-316
> > overshoots 306
> > for thirsty stock 162-163
waterholes 33, 75, 87, 93, 94, 98, 240, 251, 304, 306, 307
Wave Hill Station 28, 108, 133, 136, 241

Waybill of Travelling Stock *106*, 107
weather 27, 228, 229, 273, 304, 305
> bad seasons 307
> cyclones 287, 288
> drought 24, 126, 228, 304, 305, 309
> floods 35, 102, 277, 286, 306
> hot 37, 70, 71, 163, 166, 229, 238, 240, 247, 287, 288, 289
> problems 74, 103, 104, 105, 304-306
> rain 74, 90, 103, 306, 307, 309
> wet 19, 103, 105, 305
> *see also* dry season; wet season
wells 94, 95, 308, 309, 311
> artesian 309

Western Australia 78, 107, 168, 178, 179, 226
    explorations 22, 188
    homesteads 285
    saddlery firm 135
    stock routes 81, 94
wet season 28, 32, 35, 145, 173, 183, 240, 247, 277, 280, 301, 307, 330
    mustering 27-28
wet years 305
wheelwrighting 247, 248
whips 12, 33, 95, 96, 202, 236, 237, 238, 267, 320, 322-324
white ants 281, 282, 291, 294
    timbers resistant to 282
White, E D 128
Wickham, Captain 21

Wieneke, John Julius (Jack) 184-187
wild cattle 16, 17, 18, 32, 37-39, 44-47, 52, 291
wild horses 131, 199, 200
Williams, R M 10, 134, 184, 186
Wilson, Reg 28, 35, 38, 54, 128, 133, 136, 151, 299, 303, 323
wood 193, 194, 281, 282, 287, 289, 294, 295
    charcoal 324
    see also timber
wooden homes 274, 287
woody weed 303
World War I 128, 130, 135, 141, 211, 226
World War II 78, 112, 117, 170, 286
Worthington, Harold 187